AF809943

Hybrid Microelectronic Materials

Related titles published by the American Ceramic Society:

Role of Ceramics in Advanced Electrochemical Systems (Ceramic Transactions Volume 65)
Edited by Prashant N. Kumta, Gregory S. Rohrer, and U. Balachandran
© 1996, ISBN 1-57498-009-2

Crystal Growth of Novel Electronic Materials (Ceramic Transactions Volume 60)
Edited by R.K. Pandey and Ruyan Guo
© 1995, ISBN 1-57498-003-3

Materials and Processes for Wireless Communications (Ceramic Transactions Volume 53)
Edited by T. Negas and H. Ling
© 1995, ISBN 0-944904-94-7

Magnetic Ceramics (Ceramic Transactions Volume 47)
Edited by Bhaskar B. Ghate and John J. Simmins
© 1995, ISBN 0-944904-81-5

Ferroic Materials: Design, Preparation, and Characteristics (Ceramic Transactions Volume 43)
Edited by A.S. Bhalla, K.M. Nair, I.K. Lloyd, H. Yanagida, and D.A. Payne
© 1994, ISBN 0-944904-77-7

Materials in Microelectronic and Optoelectronic Packaging (Ceramic Transactions, Volume 33)
Edited by Hung C. Ling, Koichi Niwa, and Vishwa N. Shukla
© 1993, ISBN 0-944904-63-7

Dielectric Ceramics: Processing, Properties, and Applications (Ceramic Transactions, Volume 32)
Edited by K.M. Nair, J.P. Guha, and A. Okamoto
© 1993, ISBN 0-944904-60-2

For information on ordering titles published by the American Ceramic Society, or to request a publications catalog, please call 614-890-4700, or write to Customer Service Department, 735 Ceramic Place, Westerville, OH 43081.

Hybrid Microelectronic Materials

▼▲▼

Edited by
K.M. Nair
DuPont Electronic Materials
V.N. Shukla
Texas Instruments

▲▼▲

Ceramic Transactions
Volume 68

Published by
The American Ceramic Society
735 Ceramic Place
Westerville, Ohio 43081

Proceedings of the Symposium on Science and Technology of Microelectronics, held at the joint ACers Electronics Divsion/ISHM Fall Meeting, November 15–17, 1994, in Boston.

Copyright © 1996, The American Ceramic Society. All rights reserved.

No part of this book may be reproduced, stored in a retrieval system, or transmitted in any form or by any means, electronic, mechanical, photocopying, microfilming, recording, or otherwise, without written permission from the publisher.

Permission to photocopy for personal or internal use beyond the limits of Sections 107 and 108 of the U.S. Copyright Law is granted by the American Ceramic Society, provided that the base fee of US$5.00 per copy, plus US$.50 per page, is paid directly to the Copyright Clearance Center, 222 Rosewood Dr., Danvers, MA 01923, USA. The fee code for users of the Transactional Reporting Service for Ceramic Transactions Volume 64 is 1-57498-013-0/96 $5.00+$.50. This consent does not extend to other kinds of copying, such as copying for general distribution, for advertising or promotional purposes, or for creating new collective works. Requests for special photocopying permission and reprint requests should be directed to the Director of Publications, The American Ceramic Society, 735 Ceramic Place, Westerville OH 43081, USA.

For information on ordering titles published by the American Ceramic Society, or to request a publications catalog, please call 614-890-4700.

Printed in the United States of America.

1 2 3 4–00 99 98 97 96

ISSN 1042-1122
ISBN 1-57498-013-0

Contents

Preface

A hybrid microelectronic circuit is considered to be a micro-miniaturized modular electronic block containing various functional units, which can be integrated with semiconductor integrated circuits. The complex hybrid circuitry contains both thick and thin film–based printed simple hybrids and component packages such as resistors, capacitors, transistors, and inductors. The major advantages of the hybrid incorporated microelectronic systems are that they are flexible and therefore can be incorporated into integrated circuits to satisfy specific functions, they are small in size and hermetically reliable, they can incorporate mixed technologies without much difficulty, and they can be produced cost-effectively for almost all applications.

Hybrid microcircuits have penetrated all major industrial electronics markets, such as computers, radar and communications equipment, satellites, aircraft and navigational equipment, military electronics, medical equipment, automotive devices, and consumer electronic devices.

With the drive of new initiatives in the areas of advanced materials and technology, it is highly desirable to review and project the direction in hybrid microelectronic materials design, preparation, and other aspects of this unique technology. The decision of the American Ceramic Society and the International Society of Hybrid Microelectronics to organize jointly the First International Symposium on Science and Technology of Hybrid Microelectronics gave an opportunity to review some topics in depth by bringing together leading researchers and device engineers from universities and industry.

This volume is a collection of selected papers from the symposium and other individual papers. All the papers were reviewed through a peer review process. We acknowledge and appreciate the contributions of the speakers, session chairs, manuscript reviewers, and American Ceramic Society officials for making this endeavor a successful one.

K.M. Nair
V.N. Shukla

Telestic (Purposeful) and Atelestic (Curiosity-Driven) Research in the Post Supercollider World

Rustum Roy
Intercollege Materials Research Laboratory
The Pennsylvania State University
University Park, PA 16802

ABSTRACT

The idea that science leads "inevitably" to technology has been at the base of the disastrous state of U.S. technology and government funding of R/D. This absurd concept dominated national and industrial policy during the period bracketed symbolically by July 17, 1945 (Alamogordo), to October 21, 1993 (Washington, U.S. House of Representatives), when the SSC was defeated. Since then things have changed dramatically in R/D policy in the private and public sector. In the new post-SSC era, U.S. (and Western) industry has realized its error in its R/D posture and shut down corporate "basic" research not connected to a product line. In the public R/D policy the Clinton administration was focusing on "strategic" research and on technology policy including funding civilian product development. Although initial Congressional forces appear to be different, it is virtually certain that esoteric basic research will be steadily cut back to very much lower levels.

To the extent authorized under the laws of the United States of America, all copyright interests in this publication are the property of The American Ceramic Society. Any duplication, reproduction, or republication of this publication or any part thereof, without the express written consent of The American Ceramic Society or fee paid to the Copyright Clearance Center, is prohibited.

INTRODUCTION

In 1985 Deborah Shapley and I published the first book—*Lost at the Frontier*[1]—which dared to criticize the entire basis of U.S. science as it had evolved in the post WW II era. We claimed, on the basis of personal observations and first-hand experience, of all agencies, within a wide circle of scientists and laboratories and agencies, that one reason for the decline of U.S. technological prowess was a fundamental error in policy. We developed this theme at some length in the book,[1] to which the reader is referred for details, but which may be summarized as follows.

The U.S. scientific community—uniquely in the world—had for reasons developed in my earlier papers[2-5] come up with a completely idiosyncratic view of the relationship between science and technology which is encapsulated perfectly by two quotes from highly influential representative figures in U.S. R/D policy making:

> "He was convinced ... that 'science leads inevitably to innovative technology, and thus to prosperity.'" G. A. Keyworth, *Physics Today,* April 1985.
>
> "This is the way it works: Science → Applied Science → Technology." N. Hackerman, Former Chairman, National Science Board, *Technology and Society,* 1986.

The trajectory of the costs of particle physics and astronomy research could have convinced a student with high school algebra that it was unsustainable. But the religious worship of science among American politicians and bureaucrats, exceeded in history by only one other group, the Communist High Command and its apparatchiks in the Kremlin, was such that it had continued to spend "taxpayers hard earned dollars" on the most esoteric of pursuits for 50 years. Well may one ask: Where was the now sacred "marketplace" in such decisions?

But all this finally caught up with the country on October 21, 1993, when in a decisive 2:1 vote the U.S. House of Representatives convincingly defunded the SSC. In that one act the U.S. and the world had torn up the old contract of society with science and started to reach out for a new relationship. That such a preposterous proposition could have persisted and be proclaimed excathedra into the mid-eighties (and that it is still believed in by perhaps the majority of the scientific community) shows the extent to which scientists have disocciated from the rest of society. It is not that society had changed towards being more ignorant about science, it is that scientists have redefined themselves into an esoteric position not corresponding to reality within human society.

In Figures 1 and 2, I present our[1] schematic—which is now widely copied all over the world—of the relation of science to technology. Any attempt to do better at research policy or education policy will need to clearly define this relationship and to attempt more precise definitions of what one means by "applied," "basic," "strategic," etc., research. I urge readers who wish to enter the debate to spend the time in reading the literature on the history of technology and of science.

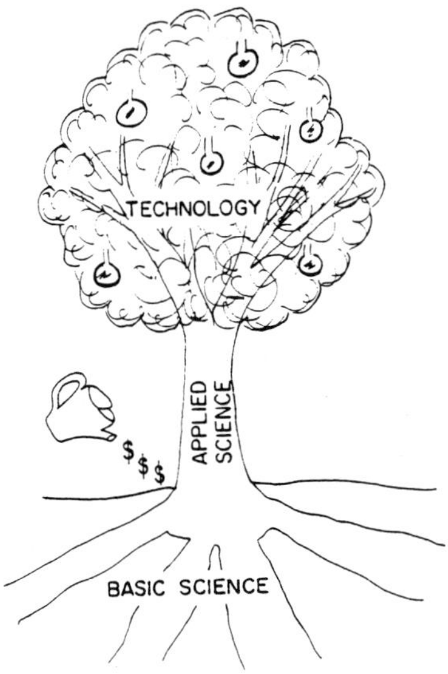

Figure 1 The conventional wisdom about the relationship between science and technology is represented as a single tree. If its roots, which are basic science, are watered, then the fruits, which are technology, will grow automatically. This widespread belief is disproved by the postwar experience of Great Britain (which has watered its basic science roots but has difficulty growing technology) and Japan (which has little basic science but grows technological fruits).

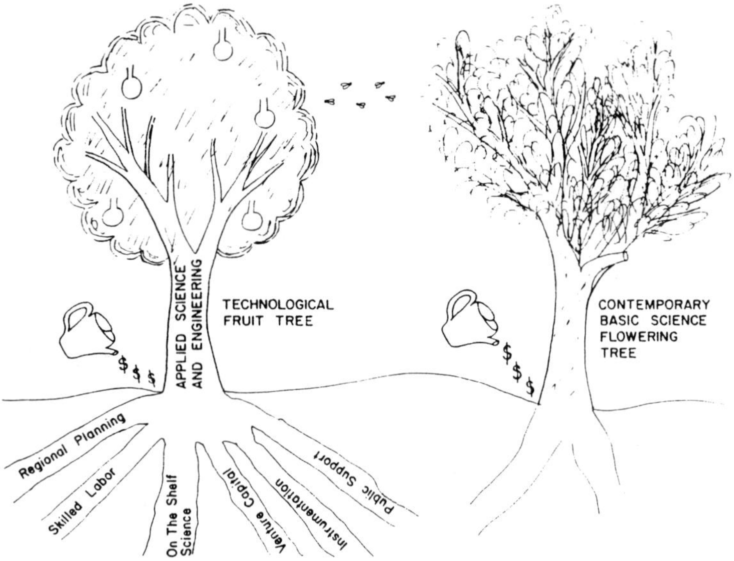

Figure 2 A more accurate metaphor is two trees, one for basic science and one for technology. This recognizes their distinct intrinsic character and the fact that they are nurtured separately by separate policies. Each tree, properly nourished, produces its own fruit: the basic science tree grows Nobel Prizes, and the more complex applied science-engineering-and-technology tree grows technology. Since the trees depend on each other as well, through bees that cross-fertilize, this model emphasizes the importance of establishing policies that nurture not only the technology tree in its own right, but the linkages between the two.

THE WATERSHED OF THE NINETIES

What happened in the nineties was that some half dozen major industrial CEO's and R/D chiefs realized that the typical

corporate lab had no value at all to the corporation in terms of return on investment (ROI). That did *not* mean there was no role for research. Quite the opposite. Research "pulled" by corporate product needs or process developments, continued as vigorous as ever. And that is the only way it makes economic sense. I have coined the term "Technology Traction" for this "sustainable R/D" pattern. I also have distinguished clearly between different kinds of science. S^1 = science for science's sake playing by the community's role; S^2 = science *for* society (public or private); and S^3 = serendipitous science. Figure 3 helps to explain these terms:

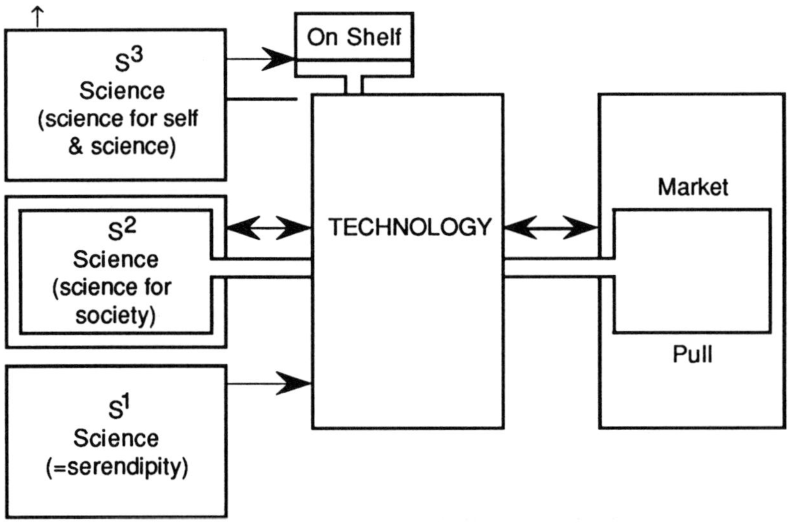

Figure 4 shows from a recent *NY Times* story how drastic has been the downsizing of research in major American companies. Now that the captains of industry have awoken to the fact that "curiosity" research does not pay returns to the specific investor in the research, how long can it be till the Congress and other representatives of the public sector realize the same thing. This need not be at all "bad" for curiosity science. What I am arguing for in science funding is so far from what most scientists believe that they often mis-state my position. Since I have the advantage of having read the literature for 25 years and written some of it, I want to put down as succinctly as possible the outlines of my position:

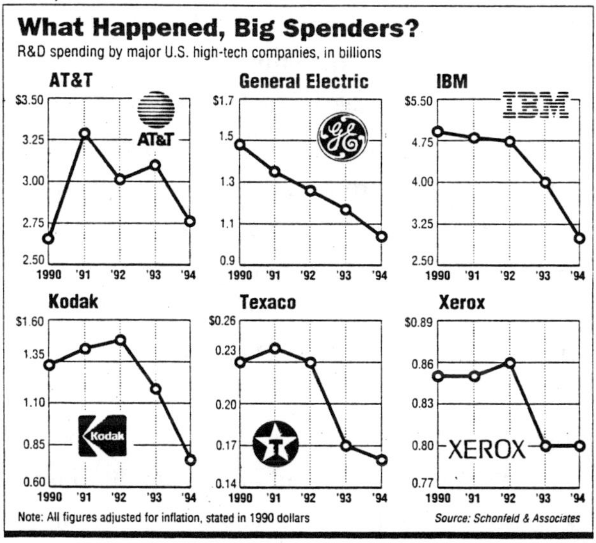

Figure 4. *Wall Street Journal* report on decline of R/D spending by major U.S. corporations, showing decline of up to 40% in four years.

1. Atelestic (i.e. "without purpose," "really basic," see Ref. 1) research *cannot* and should not be funded, by public sources, at levels beyond those for the Arts and Humanities. There should be no *proposals* written, to any agency, for such atelestic (basic) research. Indeed it is obviously an oxymoron to write a proposal describing the purpose of "without purpose" research. Thus the "basic" research parts of agencies such as NSF and DOE-OBES should either be terminated or produce both another rationale and a new process for disbursing their funds. History is on my side. The last of the great basic discoveries of physics—quantum mechanics now 70 years old—like its predecessors for 100 years was not a funded "project" of anybody's. The really fundamental work is always a by-product. In the case of quantum mechanics it was a by-product of the normal professorial role of Born, Heisenberg, Schrodinger, Dirac, etc. They were paid to teach, to profess and, as part of that, they did this research, WITH NO CONSTRAINTS, "on the side."
2. As originally suggested by Alvin Weinberg, I propose that all atelestic research should be an overhead on other work for society by the P.I. This may be (a) teaching or (b) telestic ("with purpose" or applied) research on problems which either the public or the private sector will pay for. This would mean that *most* scientists active in research would be involved in some telestic (applied)

research and thereby earn the right to do *with zero constraints*, some atelestic work. Whatever that individual thinks could be most interesting work. As a result the somewhat diminished access to complex, scope-narrowing, expensive equipment would be more than offset by being able to target the really high-risk problems, not just what is in fashion or what those twin thought-police—the so-called peers and the agency managers—prefer.

3. Most of the major discoveries in the difficult complex fields (like materials and medicine) come via serendipity. An experienced perceptive investigator spots the anomaly (see my algorithm in Ref. 5, p. 2300): like Sir Alexander Fleming did with penicillin, or their respective discoverers did with lithium as psychopharmacological agent, with glass-ceramics, with teflon, with polyethylene, with ceramic superconductors, etc. Novices and beginners have seldom made any contribution in such fields, where careful empirical work modelled on that by giants like Edison, has demonstrated time and again, its decisive superiority over assorted calculations from selected "first principles" and computer modeling. Twenty years of active research with the incredible computer capabilities now available have failed to yield a single successful prediction of any new material, leave alone its actual production. This background in the history of scientific discovery means that all research

funding should be based on *only* one parameter: prior performance. All research agencies need to do is to devise quantitative and personal, qualitative schemes for assessing a P.I.'s track record or performance. I have developed this theme in detail elsewhere[6,7] The Bible says "You do not gather grapes from a bramble bush." In research by far the best predictor of success is prior (recent) track record, not a proposal. For years I have advocated that every nation should reward its *most* productive scientists with simplified research awards based on NO proposal whatsoever.

ALTERNATIVES TO RESEARCH

This is the title under which William R. Hewlett, no anti-science Luddite, points out that for the private sector there are many routes to innovation which involve no (basic) research. For many companies the ratio called "$\frac{\text{search}}{\text{research}}$" has become important. Doing one's own research is both inordinately expensive and very slow. Searching diligently, at one tenth or one hundredth the cost, may give you not the same (useless or) useful new knowledge, but can often give you as much new knowledge. "Searching" involves today a systematic use of all the sources of knowledge: using sophisticated search routines such as the ISI uses on various databases, the older invaluable compilations of reviews and textbooks, visits to the most important centers, trading of patents, and most importantly, (telephone) discussions with the most knowledgeable experts. A moment's reflection would tell the reader that the same money and time invested in "Search" is bound to be more productive than starting up research, except where something really new is essential. Only in the rare (<1%) case would the "Research" have paid off in a significant discovery. And aye, but here's the rub—even with successful *research*, there is no way to assure that the new discovery may not pay off in products for someone other than the group that paid for the research, perhaps even in India or Taiwan or Korea. Since (see Figure 2) so many different elements are essential for the technology tree to prosper, the country of origin of the discovery, may not have the other elements and be left in the cold when the value was added by the synergesis from the other elements.

EDUCATION OF THE PUBLIC ABOUT SCIENCE AND SCIENTISTS ABOUT THE PUBLIC

I am one of the very few active research scientists who has worked for 25 years on educating the public about technology. Many research studies especially by Jon Miller of Northern Illinois University document the scientific and technological illiteracy of the public and scientists are wont to rail about this. But I have found another illiteracy equal prevalent. That is illiteracy about politics and citizenship on the part of the scientist-community. The question facing the scientist who gets into educating some group is this: what group will you focus on?

1. The specialist in your discipline.
2. The general engineer, to be introduced to your discipline.
3. The general undergraduate (90% of all students).
4. The general (attentive) public.

I believe emphasis today on Group 1 is a poor use of time because so many professors work at it full time all the time. Every university knows best, and will accept from outside a *small* amount of useful new *materials,* but never advice on curriculum matters. Group 2 has also been worked on extensively, but less than Group I. Group 3 is, in my view, the major opportunity, say, for materials scientists and engineers to insist that *every* undergraduate be exposed to *real* science, i.e. that which can be touched or felt or experienced, as contrasted to the abstract sciences of physics, chemistry and biology. Of all the "real" sciences, materials is the ideal one through which to communicate science. The facts are in: the present system of requiring all American college students to take (3 or) 4 courses in *"science"* has given us that illiterate mass of citizens. This approach can *never* work because the students have zero *interest* in "science" of that kind—not because they are not smart enough. There is an alternative: eliminate physics, chemistry and biology—PCB!—from the K-12 and nonspecialist college students choice as the representative subjects called "science." Give them *real* science instead. "Real" science is what the adjective says: That which you can contact via your senses—sens-able science. It is the way all of science started—explaining the observable and sens-able world around us. Most humans are curious about that, and that has motivated much of science. *Food* and agriculture; the *earth; materials* for shelter, clothing, transportation; everything for their *health; engineering* for the human-made world around them. In this kind of learning the materials community has the greatest opportunity and the greatest responsibility. We are the community that controls the highway of *interest* and understanding of real science. The logic of learning is simple: Start with societal problems, real objects, concrete reality, move on to the relevant technology (and only then to scientific principles, if at all). Human need, interest and curiosity meet daily in the myriad materials each of us encounters every day. (See Ref. 8 for development of this theme.)

CONCLUSION

Within the national S/T community, I believe the academic world of science and engineering has been more shielded than any other from the major changes in our society. It is urgent that this neglect of the socio-political reality around us be made up immediately. It will not be easy because it demands reading an unfamiliar literature even while the pace of change accelerates. Industrial leaders and scientists need to "educate" their faculty colleagues on the new realities in the world around. Within the technical community the materials scientists and engineers have a special opportunity to model the new relation of science to society.

REFERENCES

1. Deborah Shapley and Rustum Roy, *Lost at the Frontier,* ISI Press, 1985.
2. Rustum Roy, "The New Materials/Market Development Cycle," Inst. of Metals, London, 1995, pp. 3-28.
3. Rustum Roy, "New Approach to K-12—"Materials" as Part of Applied Science and Technology," *J. Mater. Educ.* 12:1-15 (1990).
4. Rustum Roy, "Materials Technology—A Servant of Humankind?" *Proc., TeknoVision Conference '92,* Copenhagen, 1992, pp. 77-90.
5. (a) "Science, Technology and Social Problems," *Ceram. Eng. Sci. Proc.* 12:2275-2311 (1991).

   (b) "The Nature and Nurture of Technological Health," *Ceramics and Civilization, High-Technology Ceramics—Past, Present, and Future,* W. D. Kingery (ed.), Am. Ceram. Soc., 11:351-370 (1987).
6. Rustum Roy, "Alternatives to Review by Peers: A Contribution to the Theory of Scientific Choice," *Minerva* 22(3-4):316-328 (1985).
7. Rustum Roy, "Funding Science: The Real Defects of Peer Review and An Alternative to It," *Science, Technology & Human Values* 10(3):73-81 (1985).
8. Rustum Roy, "K-12 Education: A Primer for MaterialsResearchers," *MRS Bulletin* 17(3):5-9 (1992).

FERROELECTRIC CERAMICS: MATERIALS AND APPLICATION ISSUES

L. Eric Cross
Evan Pugh Professor of Electrical Engineering
Materials Research Laboratory
The Pennsylvania State University
University Park, PA 16802

ABSTRACT

The unusual properties of the perovskite structure ferroelectric ceramics which makes them attractive for several different fields of application include:

- The very high dielectric permittivity for capacitor dielectrics, still the largest market segment.

- The polability to high remanent polarization inducing very strong direct and converse piezoelectric effects for electromechanical transduction in sensors and actuators.

- Large pyroelectric effects in the poled structures with interesting application in bolometric infra-red detection and imaging.

- The transparency of the high band gap structure combined with very strong electro-optic effects beginning to yield new possibilities for photorefractive high density optical information storage and phase matchable broad band optical second harmonic generation.

New developments in each of these areas will be highlighted focusing primarily upon the materials issues. With our enhanced capability to deposit well controlled thin films of many of the most interesting composition families the new possibilities which these afford will be briefly reviewed.

To the extent authorized under the laws of the United States of America, all copyright interests in this publication are the property of The American Ceramic Society. Any duplication, reproduction, or republication of this publication or any part thereof, without the express written consent of The American Ceramic Society or fee paid to the Copyright Clearance Center, is prohibited.

The film structures provide new capability to generate exceedingly fine controlled microstructure which raises new concerns as to the influence of scale upon the cooperative phenomena which give rise to both ferroelectric and antiferroelectric phases. Experience with the self assembling self limiting nanostructures which appear in relaxor ferroelectrics and their influence on properties will be briefly reviewed.

INTRODUCTION

In considering the full family of applications for ferroelectric ceramics in dielectric, piezoelectric, pyroelectric, and electro-optic systems it is evident that the properties exploited are softness, thus unlike structural ceramics where it is the elastic stiffness which is of most interest, in functional ceramics it is the control and exploitation of compliance which is important [1].

In conventional dielectric systems the multilayer ceramic capacitor (MLC) is the most widely used component with a market approaching 100 billion units worldwide per year [2]. Tape casting is widely used for assembly and the dielectrics are mostly barium titanate based. It will be shown that the higher permittivity units draw upon the very high intrinsic polarizability near the phase transitions. Newer lead based relaxor ferroelectric systems will be examined to highlight the rather different spin glass like response. However, even though these composition can be co-fired with both silver and copper metallurgy they are not yet a major segment of the MLC market. As the voltage of semiconductor electronics decreases, dielectric layers must get thinner, exploration of recent techniques for forming submicron thin films of the lead based ferroelectrics suggest that these may offer interesting new capacitance structures for high density semiconductor packaging.

Piezoelectric ferroelectrics for sensor and actuator functions in mechanical systems are a rapidly developing market, where the lead zirconate:titanate (PZT) compositions are very widely used. The behavior in these poled ceramic elements is very complex and even after more than 40 years of use, the balance of contributions to response, particularly in the 'soft' high sensitivity compositions is still not fully understood. For sensing, composite systems give the possibility to 'tune' the tensor response and offer new possibilities in both large area and high frequency acoustic systems. Piezoceramic actuators can supply immense mechanical force, but only over minute throw. Simple mechanical amplification as in flextensional, moonie and rainbow structures are of increasing interest. Energy density consideration in thin film systems suggest that ferroelectric piezoelectric films must be important new players in micro-electro mechanical systems (MEMS). Piezoelectric diaphragm type micro motors give high torque at low speeds with reliability now up to 10,000 hours of continuous operation.

Pyroelectric point detectors are extensively used for thermal sensing in a wide range of industrial and consumer applications. Current research focus is upon long wavelength IR imaging where both hybrid and thin films compete. Very low thermal mass with high thermal isolation of the pixel plane is essential for high sensitivity, and again thin films are of special interest in new air bridge isolated imagers.

Optical applications of ferroelectrics mostly draw on the high electro-optic and non-linear optic characteristics of ferroelectrics. Guided wave structures on single crystal lithium niobate are a mature technology using indiffusion to modify the index. Recently stress optic modification has been demonstrated in strontium barium niobate which may permit more complex structures with lower attenuation. In harmonic generation quasi phase matching using domain divided crystals promises high intensity blue light from laser diodes. Perhaps the area which shows most promise is that exploiting the strong photo-refractive effects in $BaTiO_3$, $KNbO_3$, and $Sr_{0.6}Ba_{0.4}Nb_2O_6$. Optical amplification, frequency conversion, information storage, pattern recognition and optical soliton generation have all been observed.

In the following article, the general principles controlling properties of interest will be briefly reviewed, the wide range of information available will be referenced and the focus will be upon new developments. For each area, the impact of the growing availability of high quality ferroelectric thin films will be assessed.

GENERAL CONSIDERATIONS

Ferroelectric crystals are named for the analogy of their dielectric properties to the magnetic properties of ferromagnets, however a number of important qualitative and quantitative differences strongly impact the properties and the manner in which they can be exploited in devices.

1. Unlike magnetism, electric monopoles (carriers of charge) are common in all systems. The resulting charge transport properties strongly influence the static and low frequency dynamic dielectric, piezoelectric, pyroelectric, and electro-optic properties. There is always a decay of depolarizing fields and a strong influence on the electrical boundary conditions in ceramics.
2. Electrical ordering is soft so that although many systems follow Curie Weiss behavior in the para form i.e., The susceptibility η is given by

$$\eta = C/CT - \theta) \qquad (1)$$

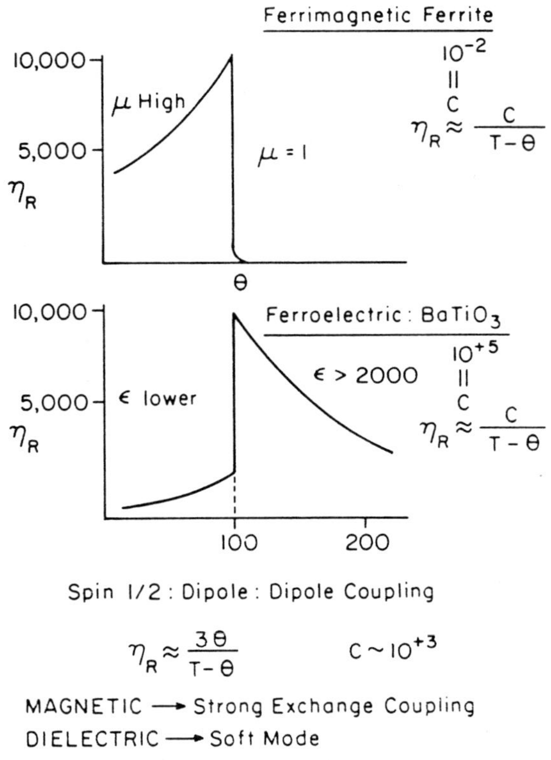

Figure 1 Comparing Curie Weiss behavior in a $BaTiO_3$ type ferroelectric with the Curie Weiss behavior of a typical ferrimagnetic ferrite.

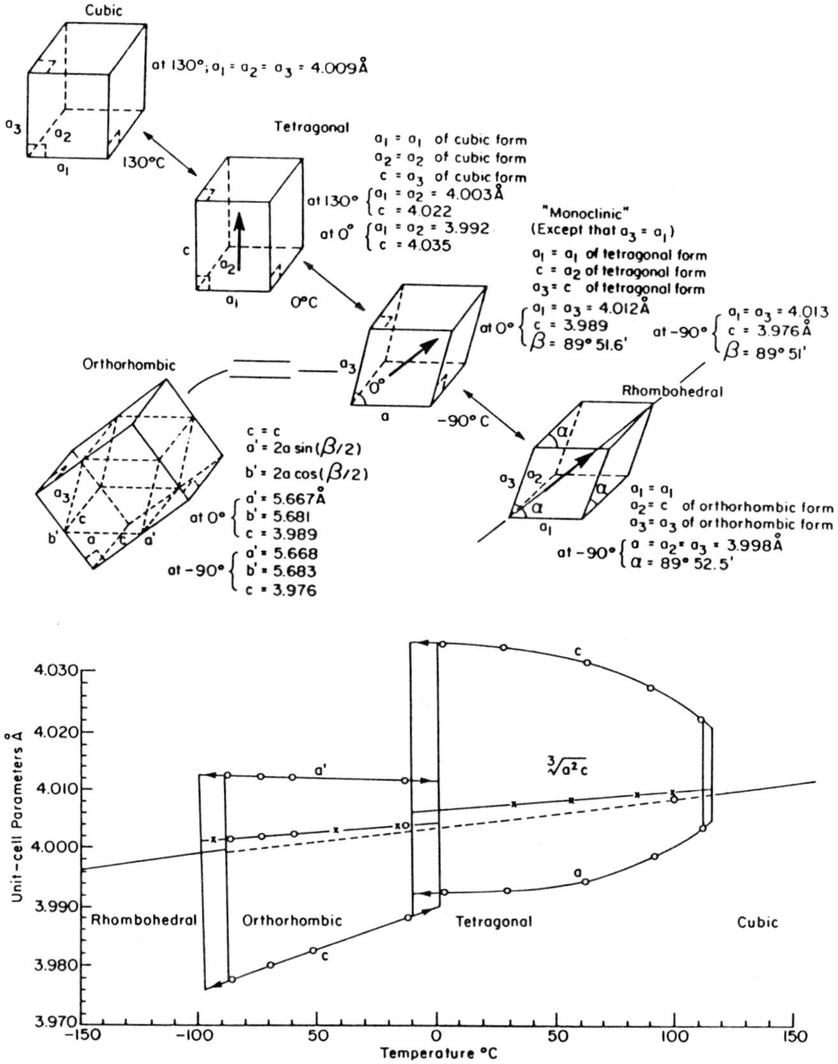

Figure 2 Strains associated with the spontaneous polarization in the ferroelectric phases of $BaTiO_3$.

where C is the Curie Constant and θ the Curie Weiss temperature. The ferroelectric C ~ 10^{+5}K compared to the magnetic C ~ 10^{-2}, very high dielectric permittivity persists in the paraelectric phase of the ferroelectric for a wide range of temperatures above θ (fig. 1) and even in the ferroelectric phase the state of ordering can be quite markedly affected by rather modest field levels.

3. Since the polarization in the oxide ferroelectrics is mostly generated by the displacive relative motions of ions in the structure it is not surprising that the coupling of polarization to the lattice is much stronger in the ferroelectric. Spontaneous deformations on polarizing are often of a few % (fig. 2) and elastic energy is a strong contributor to the nature of the Curie point transition. The high induced strain suppresses fluctuations in polarization and leads to whole families of useful electrostrictive devices.
4. The oxide ferroelectric of interest are high band gap insulators with good optical transparency in visible light and large electro-optic and non linear optic constants. Both these families of properties involve electric field induced optical anisotropy, are difficult to exploit in transparent polycrystal ceramics and need high quality single crystals. To tune properties however as in the other property areas it is necessary to modify and control the lattice by suitable solid solution. Solid solutions are however by definition incongruently melting and both difficult and expensive to grow in large optically perfect crystals. Thus it is perhaps not surprising that in spite of many very promising device concept this area is the last to achieve full commercialization.

It does however appear probably that with the developing new techniques for growing excellent quality epitaxial oxide films for high T_c superconductors now being applied to ferroelectrics it may soon be possible to achieve quality oriented films of any solid solution composition, making possible the exploitation of many of the earlier concepts in much better optimized materials.

DIELECTRIC SYSTEMS

In randomly axed polycrystal ceramics, ferroelectrics derived from high symmetry prototypic (paraelectric) phases are essential both to avoid anisotropy above T_c and to provide many orientations for the polarization below T_c. Compositions in the perovskite structure are universally applied (fig. 3) with modified barium titanate $BaTiO_3$ much the most widely used. For very high permittivity (Z5U, Y5V) characteristics the high dielectric peak at the Curie temperature is maneuvered by solid solution into the room temperature region, and broadened by deliberately induced heterogeneity (fig. 4). Some typical formulations are given in Table 1 and characteristics of permittivity versus Temperature in figure 5 [2,3,4].

For the lower flatter (X7R) characteristics also shown in figure 4, higher purity $BaTiO_3$ is used but grain growth is inhibited. In the fine grain ceramic the inability to form a twin (domain) structure leads to a high internal stress which raises the intrinsic polarizability near the 0°C tetragonal:orthorhombic ferroelectric transition. The grain growth inhibitor can be chosen to "pad" the Curie point peak leading to a broad flatter but lower permittivity response.

TABLE I. Typical Practical $BaTiO_3$ Based Dielectric Formulation from Herbert.

Item	K_{max}	Temp. range for K_{max} -20%	AO/BO_2	Ba	Sr	Ca	Mg	Ti	Zr	Sn
					Composition in cation%					
A	2280	-20 to+100	0.98	48.5	0.73	–	–	49.5	–	0.73
B	3260	0 to 120	1.03	45.3	–	4.8	1.08	44.0	4.8	–
C	3450	-10 to 72	1.01	44.2	–	5.7	0.75	42.9	6.4	–
D	5000	12 to 75	1.02	47.6	2.5	–	0.45	45.3	0.45	3.38
E	8200	42 to 70	1.005	35.5	12.5	1.8	0.8	46.9	2.6	–

An alternative approach for very high K Z5U type dielectrics is to use the lead based relaxor ferroelectric compositions based on $PbMg_{1/3}Nb_{2/3}O_3$ (PMN). In PMN limited 1:1 ordering of Mg:Nb cations on the B site of the ABO_3 structure leads to nanoscale compositional heterogeneity [5,6] and the appearance of small micropolar regions at temperatures well above the dielectric maximum [7]. In the relaxor, electrocrystalline anisotropy is inadequate to stabilize the micropolar P_s vectors and a new large fluctuating component is added to the polarizability [8]. The dielectric maximum is due to a freezing out of the fluctuating component with the expected strong dielectric dispersion [9].

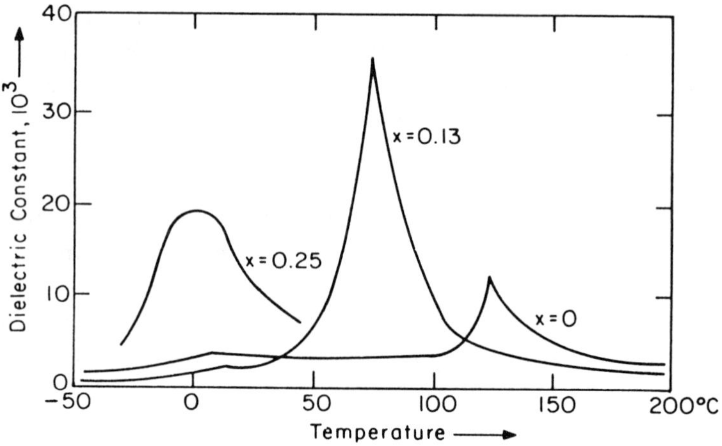

Figure 3 Permittivity temperature curves for solid solution in the $BaTiO_3$:$BaZrO_3$ composition system.

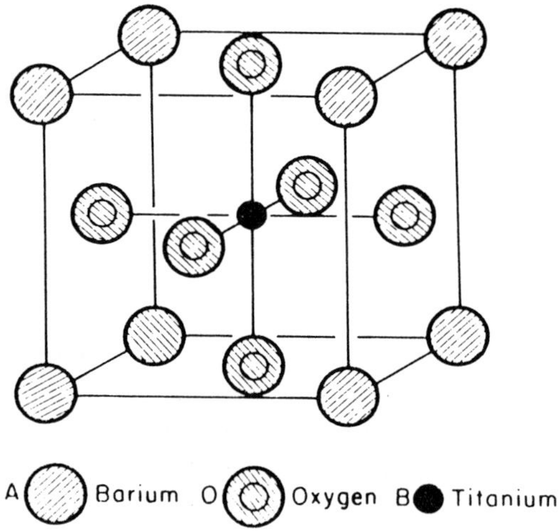

Figure 4 The unit cell for a cubic perovskite ($P_{n1}3_m$ symmetry) $BaTiO_3$ at a temperature above T_c.

randomly axed ceramic inducing conical symmetry and strong non zero piezo coefficients d_{31}, d_{33}, and d_{15}[22].

(d) In compositions close to the MPB which is near vertical on the temperature/composition diagram one may stay close to the instability of the phase transition over a very wide temperature range, with correspondingly high values of the intrinsic polarizability and deformability.

(e) One must be in the ferroelectric state with a mobile domain structure at high fields to be able to induce the symmetry breaking remanent polar state essential to induce piezoelectricty. However, as a consequence the use state is a metastable poled domain ensemble (fig. 9). It is clear from the + - symmetry of the total figure 9b that the basic deformation mechanism is electrostrictive, however for a limited range of fields about the poled working point the deformation (fig. 9b) is clearly piezoelectric.

(f) To control the piezoelectric response it is important to control the electrical coercivity. i.e. the width of the hysteresis loops. For very low drive fields low coercivity is desirable as the saturation arm becomes steeper yielding higher piezoelectric constant. In high drive systems, a very high coercivity is desirable so that the larger driving field does not induce depoling and consequent hysteresis loss and heating, but the levels of $d_{31}d_{33}$ will be lower due to the flatter saturation arm.

Even at very low drive fields we may expect the response to be complex with possible contributions to both polarization and elastic strain from intrinsic polarization change in the domain, reversible domain wall motion in both tetragonal and rhombohedral domain structures, possible reversible motion of the phase boundary separating polar tetragonal and polar rhombohedral regions in the same grain.

Intrinsic single domain response has been addressed in a series of papers applying Landau Ginsburgh Devonshire (LGD) phenomenological theory [23,24,25,26,27]. For practical application a very large range of substituents have been tried to control the extrinsic domain wall component of response. Lower level additives (fig. 10) can be divided into donor like, i.e. substituents which carry more electron charge than the ions for which they substitute in the lattice (fig. 10a) and acceptor like, i.e. substitutent which carry less charge than the ion being substituted (fig. 10b). The acceptor ions appear to pair with oxygen vacancies in the lattice to provide a slowly changing defect dipole which orients in the domain field stabilizing the

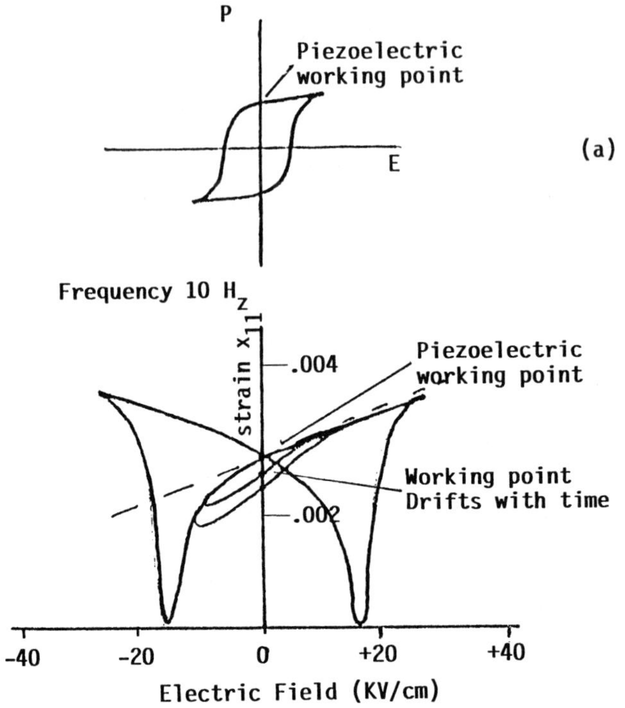

Figure 9 (a) Polarization vs. Field in a PZT ceramic.
(b) Piezoelectric/Electrostrictive strain vs. field large strain electro.

MODIFIERS OF THE EXTRINSIC RESPONSE

DONOR ADDITIVES		ACCEPTOR ADDITIVES
Nb_2O_5	$PbNb_2O_6$	Fe_2O_3
Ta_2O_5	$PbTa_2O_6$	Al_2O_3
WO_3		Cr_2O_3
Bi_2O_3		MnO_2
Sb_2O_3		MgO
La_2O_3		NiO
Wall Motion Promoters		Wall Motion Inhibitors

Figure 10 Common lower level (>5 mole%) donor dopants an acceptor dopants use in 'soft' and 'hard' PZT formulations.

existing variant at the expense of others and thus indirectly stabilizing the wall structure [28,29]. Donor dopants lower the coercivity and strongly enhance both dielectric and piezoelectric responses [30,31]. Although a number of "hand waving" arguments have been advanced there is still no clear understanding of the mechanisms involved in the donor modified systems.

Based on PZT very many third valence compensated components have been added to form three component solid solutions, all of which soften behavior, reduce the Curie temperature. Figure 11 lists just a few of the very many compositions which have been used.

The importance of the extrinsic components to the PZT response may be adduced from the fact that even in the hardest compositions at room temperature some 25% of piezoelectric response is extrinsic, whilst in the very soft donor doped compositions more than 75% of total response is extrinsic.

ELECTROSTRICTIVE CERAMICS

In section 2, the very high permittivity which is achieved in the nano-composite relaxor ferroelectrics was briefly reviewed. It was shown that these ceramics in the PMN system may be taken to large values of polarization under high electric field, but that they do not retain a remanent polar state. Since in all perovskite ferroelectrics the basic coupling to the lattice is electrostrictive it is natural to ask whether the high polarizability may be directly exploited to produce high strain by the relation

$$x_{kl} = Q_{ijkl} P_i P_j \qquad (3)$$

where x_{kl} is the induced strain.
$P_i P_j$ are components of the polarization.
Q_{ijkl} is the electrostriction constant in polarization notation.

Electrostrictive strains x_{33} and x_{11} as a function E_3 are shown for PMN and a PMN 10% PT in figure 12ab, and the quadratic nature of the strain confirmed in figure 13 [32]. In fact for PMN and PMN:PT the electrostriction constants Q_{11} and Q_{12} are essentially independent of temperature (fig. 14) [33].

For valence compensated additions the systems studied have included:

$Pb(Ni_{1/3}Nb_{2/3})O_3:PbTiO_3:PbZrO_3$
$Pb(Co_{1/3}Nb_{2/3})O_3:PbTiO_3:PbZrO_3$
$Pb(Mn_{1/3}Nb_{2/3})O_3:PbTiO_3:PbZrO_3$
$Pb(Cu_{1/3}Nb_{2/3})O_3:PbTiO_3:PbZrO_3$
$Pb(Mn_{1/3}Ta_{2/3})O_3:PbTiO_3:PbZrO_3$
$Pb(Zn_{1/3}Nb_{2/3})O_3:PbTiO_3:PbZrO_3$
$Pb(Ni_{1/3}Fe_{1/3}Nb_{1/3})O_3:PbTiO_3:PbZrO_3$

$Pb(Cu_{1/3}Nb_{2/3})O_3:PbTiO_3$
$Pb(Cd_{1/3}Nb_{2/3})O_3:PbTiO_3$
$Pb(Zn_{1/2}Te_{1/2})O_3:PbTiO_3$
$Pb(Mg_{1/2}Te_{1/2})O_3:PbTiO_3$

$Pb(Sb_{1/2}Nb_{1/2})O_3:PbTiO_3:PbZrO_3$
$Pb(Sn_{1/2}Nb_{1/2})O_3:PbTiO_3:PbZrO_3$
$Pb(Sb_{1/2}Nb_{1/2})O_3:PbTiO_3$
$Pb(Sn_{1/2}Nb_{1/2})O_3:PbTiO_3$

$Pb(Mg_{1/2}V_{1/2})O_3:PbTiO_3:PbZrO_3$
$Pb(Mn_{1/3}Sb_{2/3})O_3:PbTiO_3:PbZrO_3$
$Pb(Li_{1/4}Nb_{3/4})O_3:PbTiO_3:PbZrO_3$
$Pb(Sb_{1/2}Nb_{1/2})O_3:PbTiO_3:PbZrO_3$
$Pb(Fe_{1/2}Sb_{1/2})O_3:PbTiO_3:PbZrO_3$
$Pb(In_{1/2}Nb_{1/2})O_3:PbTiO_3:PbZrO_3$
$Pb(In_{1/2}Nb_{1/2})O_3:PbZrO_3:PbSnO_3:PbTiO_3$
$Pb(Mg_{1/2}V_{1/2})O_3:PbTiO_3:PbZrO_3$

$(Ag_{1/2}Bi_{1/2})TiO_3:PbZrO_3:PbTiO_3$
$(Ag_{1/2}Bi_{1/2})ZrO_3:PbTiO_3:PbZrO_3$

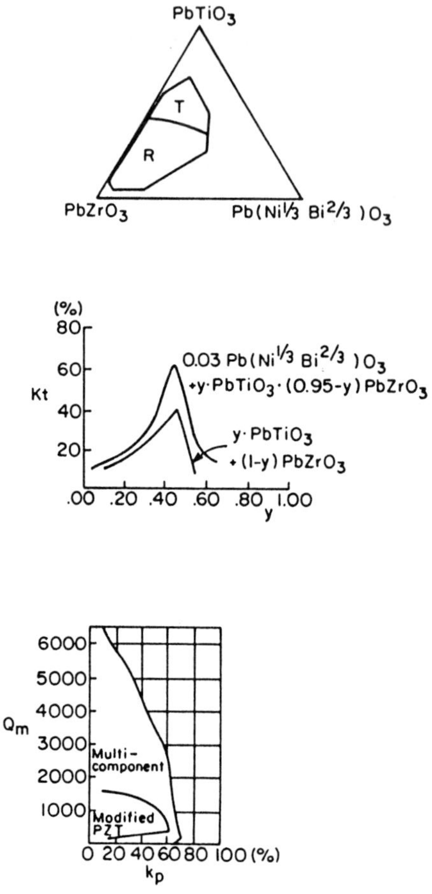

Figure 11 Valence compensated relaxor ferroelectric additives in three component PZT based systems.

An unexpected bonus for the relaxor, is that the steady accretion of polarization for temperatures above T_m leads to an expansion term in dilatation of the form

$$\Delta v \ \alpha \ (Q_{11} + 2Q_{12}) \ P_{<111>}^{2} \qquad (4)$$

which compensates for thermal contraction. Thus for a range of temperatures near 20°C it is possible to mate PMN with ULE glass diaphragms to control surface deformable mirror [34].

Because of the reproducible zero strain state in PMN, due to the absence of a stable domain structure it has found wide application in optics. Perhaps the most important recent application has been in control of the alignment tilt mirrors in the wide field and planetary camera replacement in the Hubble Space Telescope. These four small tilt mirrors permit perfect alignment of the compensators of the original mirror with a massive improvement in performance (fig. 15).

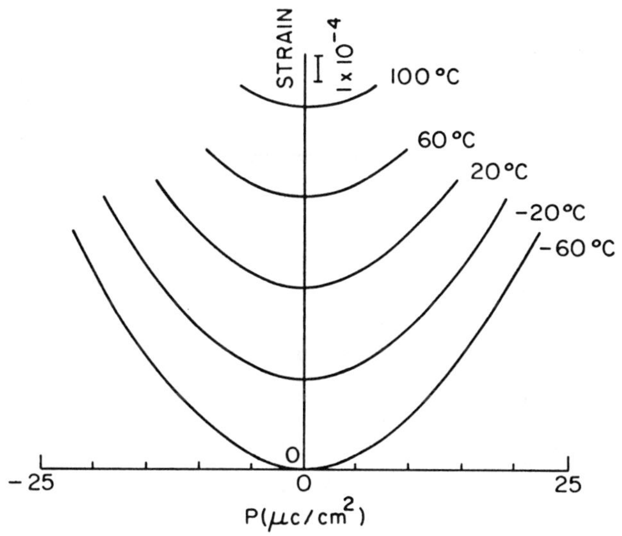

Figure 12 Typical polarization vs. strain curves in a PMN electrostrictive actuator as a function of temperature.

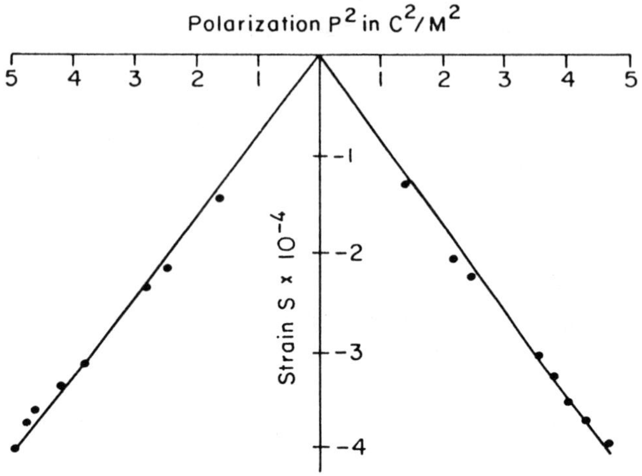

Figure 13 Quadratic electrostrictive transverse response in a PMN:10% PT actuator composition.

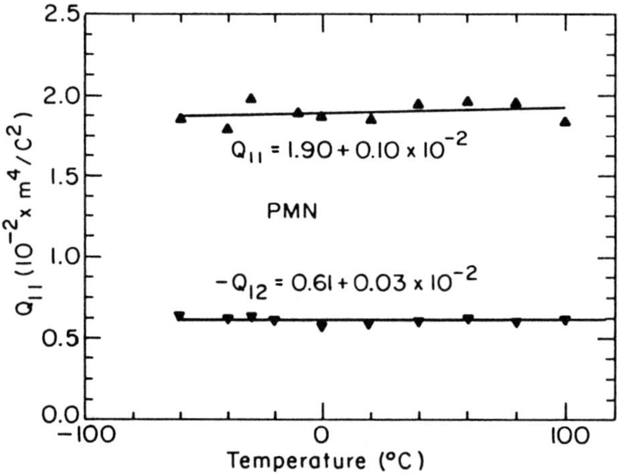

Figure 14 Electrostriction constants Q_{11} and Q_{12} as a function of temperature in a PMN:10%PT relaxor ferroelectric.

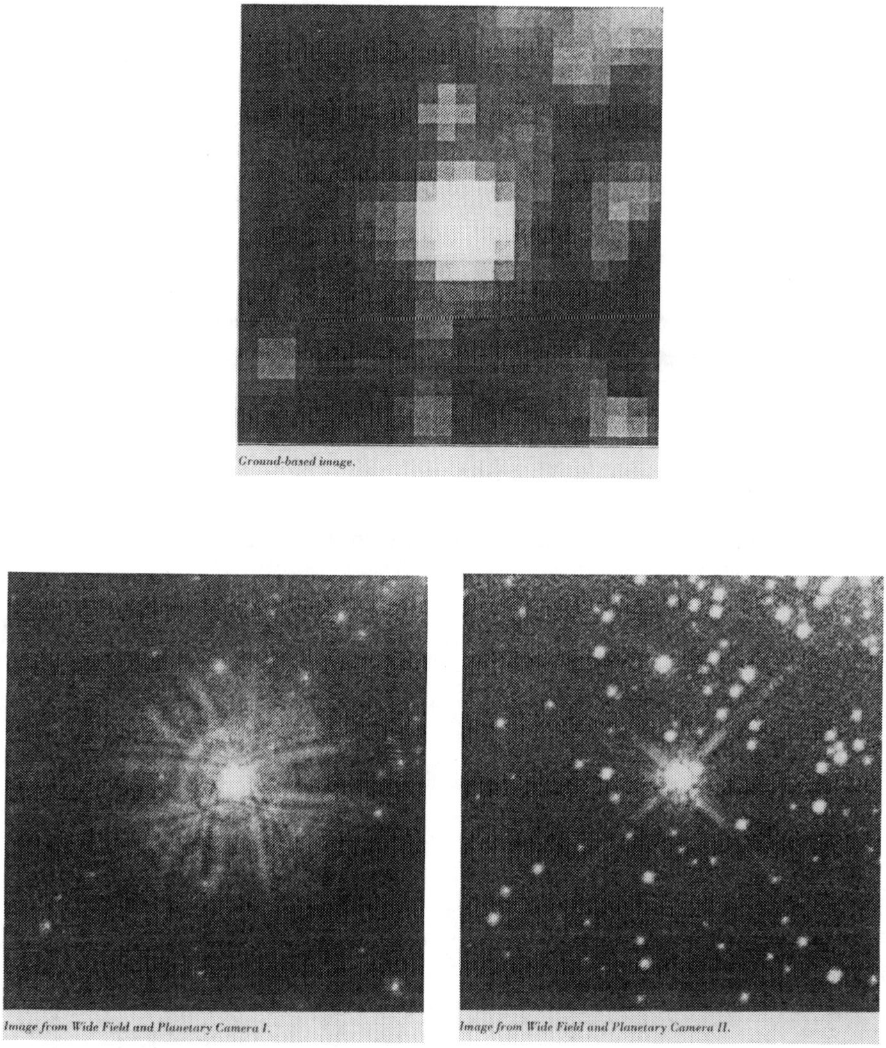

Figure 15 Example of the improvement of the image from the Hubble Space Telescope of the Star Melnick 34 using the new wide field and planetary camera II incorporating tilt mirrors controlled by multilayer PMN:PT actuators.

PIEZOELECTRIC COMPOSITES

Dielectric, piezoelectric and elastic properties of poled piezoelectric ceramics are tensor quantities and for many types of application it is possible to spell out a figure of merit for the material which often requires the enhancement of some of these tensor coefficients and the diminution of others.

A typical example is the requirement for sensing of very weak hydrostatic pressure waves using large area sensors as in many Navy hydrophone needs (fig. 16). For hydrostatic pressure (fig.17) the stress $X_{11} = X_{22} + X_{33} = -p$, so that the polarization change P_3 is given by

$$\begin{aligned} P_3 &= d_{33}(-p) + d_{31}(-p) + d_{31}(-p) \\ &= (d_{33} + 2d_{31})(-p) \\ &= d_h(-p) \end{aligned}$$

where d_h is called the hydrostatic piezoelectric charge coefficient.

The voltage generated by the hydrophone, working into a very high impedance load will be given by

$$g_h = \frac{(d_3 + 2d_{31})}{\varepsilon_{33}} \tag{5}$$

and a figure of merit often used for hydrophone materials is the product $d_h g_h$

$$d_h \, g_h = \frac{(d_{33} + 2d_{31})^2}{\varepsilon_{33}} \tag{6}$$

For high sensitivity PZTs, there is an unfortunate near cancellation such that

$$d_{33} \cong -2d_{31} \tag{7}$$

so that $d_h << d_{33}$ or d_{31}, and PZT alone is not a good hydrophone material.

In exploring composites for hydrophone applications it would be advantageous from the point of view of density and of flexibility to combine the ceramic with a dielectric polymer. Comparing the dielectric and elastic properties between two such phases, a fascinating juxtaposition is evident

Dielectrically PZT is ultra soft (k ~ 3,000) whilst the polymer is quite stiff (k ~ 10) but in the elastic response, just the converse is true. The polymer is ultra soft (s_{11} ~ $30.10^{-11}M^2/N$) but the ceramic is very stiff (s_{11} ~ $2.10^{-11}M^2/N$) thus by careful control of the mode in which each phase is self-connected in the composite one can "steer" the fluxes and fields so as to enhance wanted coefficients and diminish unwanted coefficients so as to vastly improve the figure of merit.

Over some twelve years of intensive effort to design effective composites three important basic principles have emerged.

- Connectivity. The mode of self interconnection of the phases controls the fluxes and fields in the system enabling a tailoring of the tensor coefficients.

- Symmetry. Both the symmetry of the individual component phases and the macro symmetry of their arrangement in the composite can be used for additional control.

- Scale. The model of averaging for the property coefficients depends upon the scale of the composite phases in relation to the wavelength of excitation. Unusual resonances can occur when λ and d are comparable.

A major aid in thinking about the design of connectivity was the simple cubes model [35] (fig.18) and the associated notation, now internationally accepted which describes the dimensionality of the connectivity for active and passive phases.

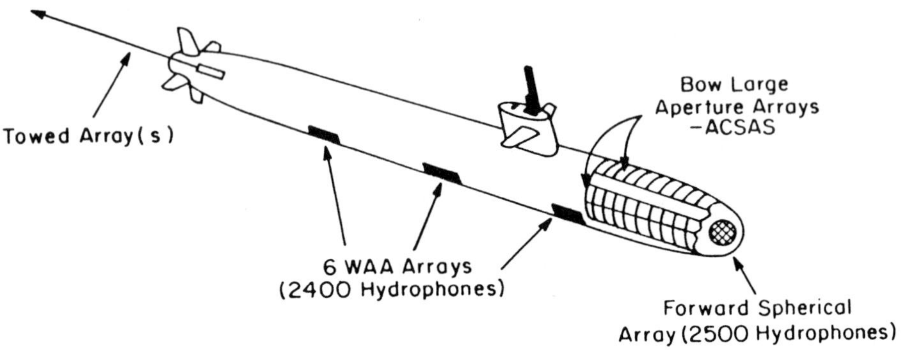

Figure 16 Examples of the need for large area hydrophone sensors in modern submarine acoustics.

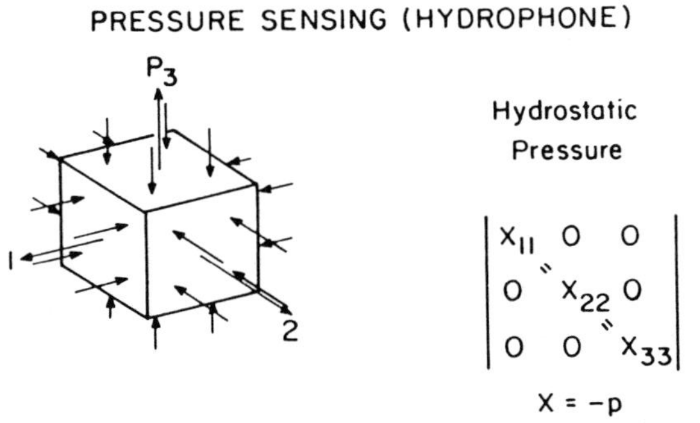

Figure 17 Stress System seen by the material under hydrostatic conditions in the hydrophone.

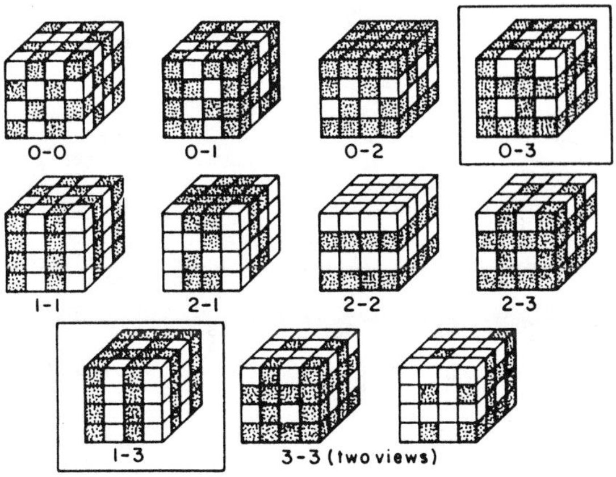

Figure 18 The simple 'cubes model for connectivity patterns possible in simple two phase piezoceramic:polymer composites.

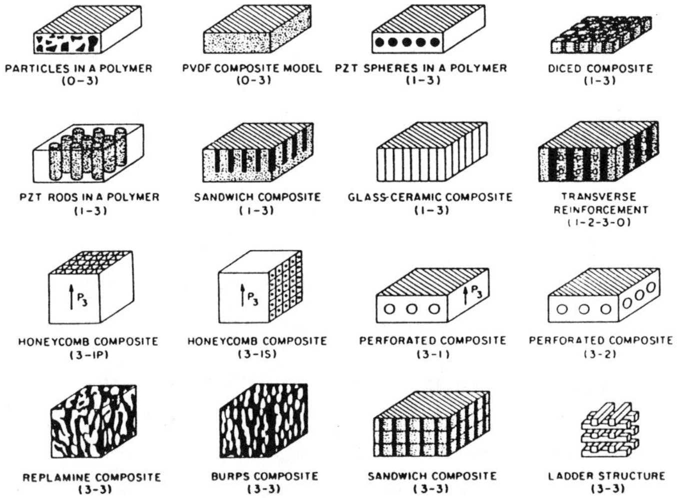

Figure 19 Examples of composite piezoceramic:polymer composites with different phase connectivities.

COMPARISON OF $d_h g_h$ OF VARIOUS COMPOSITES

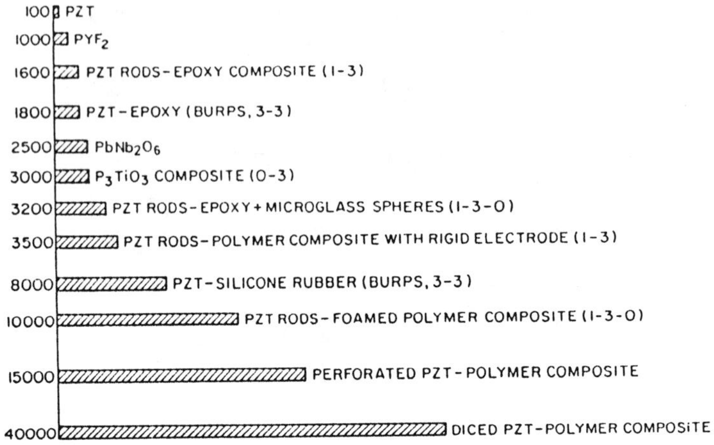

Figure 20 Hydrostatic figures of merit achieved using several different connectivities.

An indication of the many types of connectivities which have been used at Penn State for piezoceramic:polymer composites is shown in figure 19 and a measure of the improvement in hydrophone figure of merit over pure PZT for some of these systems is shown in figure 20.

The special case of the 1:2:3:0 composite which uses PZT rods in a foamed polymer matrix with transverse glass reinforcement is given in figure 21.

For the 1:3 type composites a major impediment to evolution for large scale structures has been the problem of assembly. Recently Fiber Materials of Biddeford, Maine have applied their ultraloom technology originally evolved for thick section carbon:carbon composites to this problem. Using the ultraloom they are able to stitch PZT posts into a template structure which contains the transverse glass fiber reinforcement and make sections up to four feet in width and of almost any length.

NAVY HYDROPHONE

Up to 1975 Material lead zirconate titanate piezoelectric ceramic PZT

Power figure of merit $d_h g_h$

Product of hydrostatic voltage x hydrostatic charge

$d_h g_h$ in tensor form → $\dfrac{(d_{333} + 2d_{311})^2}{\epsilon_{33}}$

$d_h g_h \sim 100 \cdot 10^{-15}$ M^2/ Newton ⇐

PROBLEM $d_{333} \cong -2d_{311}$ ϵ_{33} very large

COMPOSITE SOLUTION

TRANSVERSE REINFORCEMENT
(1-2-3-0)

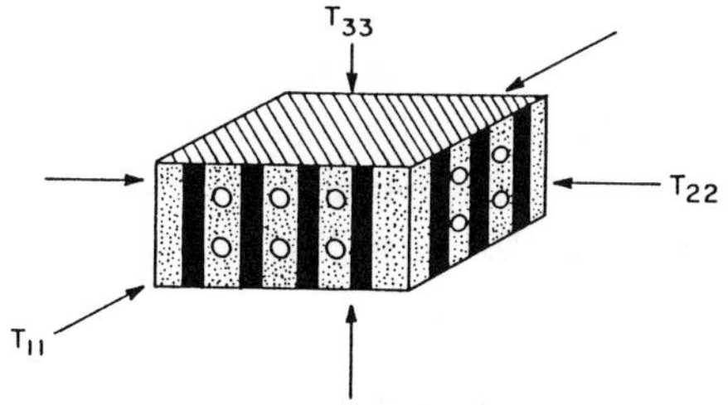

$$p = \begin{vmatrix} -T_{11} & 0 & 0 \\ 0 & -T_{22} & 0 \\ 0 & 0 & -T_{33} \end{vmatrix}$$

(rows 1–2) Taken up on transverse reinforcement

(row 3) Enhanced on PZT
Polymer acts like a tent

ϵ_{33} much reduced by the polymer
Composite 10 v/o PZT 90% rubber

$d_h g_h \cong 150{,}000 \cdot 10^{-15}$ M^2/Newton ⇐

Figure 21 Figure of merit for a 1:2:3:0 transverse reinforced foamed polymer composite vs. performance of solid PZT.

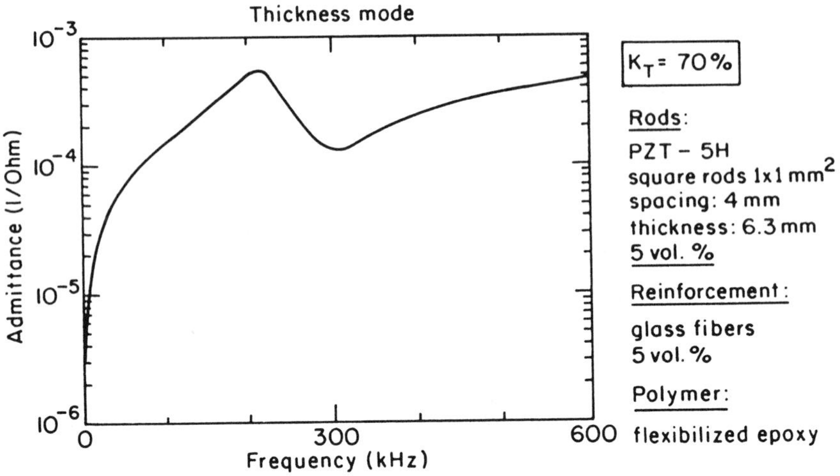

Figure 22 Thickness resonance curve for a Fiber Materials Inc. 1:3 composite containing 5 volume% PZT rods.

The FMI composites are not only interesting for very large area hydrophones, but can also be used in an actuator mode. It is interesting to note that with only 5 vol% PZT and a resultant density of 2.2 gm/cm^3 the transverse coupling coefficient k_t at 0.70 is larger than that of solid PZT (fig. 22).

The 1:3 type concept has also been applied to transducers for medical ultrasonics [36,37]. Here the required frequencies are much higher ~10Mhz so that the scale is very much smaller and the rod structure can be cut from solid PZT (fig.23). Beam characteristics, pulse shape and coupling factor are much improved over solid PZT transducers.

PYROELECTRICS

Thermal detection can be used for any wavelength of electromagnetic radiation, provided there is a convenient absorber which will convert the radiation into heat. For most parts of the electromagnetic spectrum there are other more convenient receptors, however in the long wavelength infra red for wavelength of 7 to

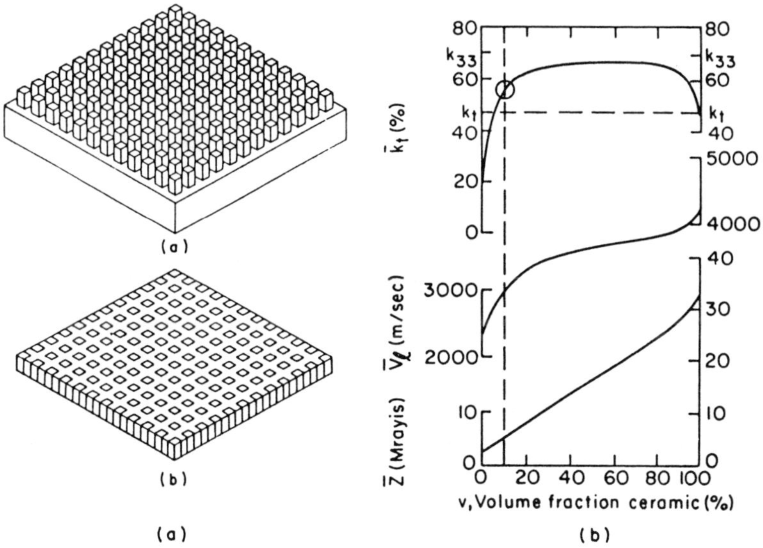

Figure 23 Piezoelectric ceramic: Polymer composites applied to electromedical transduction.
(a) The die and fill method of construction.
(b) Transverse couplinge composite as a function of volume fraction PZT.
After Smith [37].

13 μ meters photon detectors have to be cooled to helium temperatures and though highly sensitive become cumbersome and expensive. It is just in this wavelength region that detection and particularly imaging by thermal detectors becomes most interesting.

In the pyroelectric and other thermal detectors the objective is to obtain very high responsivity to longwave IR radiation so that low signals from warm bodies may be observed over background noise. In all thermal detectors the overall responsibility is the product of two terms:

(a) The induced temperature change in the detecting element per unit of incident IR irradiation.

(b) The thermometric sensitivity of the detecting element.

RESPONSIBILITY	=	Induced temperature change ΔT in the detecting element	$\times$	Thermometric sensitivity of the detector element.

Considering first the optimization of (a). Most values of ΔT for high sensitivity detectors and imagers are very small compared to the temperature drift even in a highly temperature stabilized environment so it is customary to chop the incident IR at a low frequency (10 -30 Hz) and examine only the AC component of ΔT.

For a solid state detecting element, a compact expression for (a) may be written in the form

$$\Delta T = \frac{I_o}{\left(g^2 + \omega^2 c^2 d^2\right)^{\frac{1}{2}}} \tag{8}$$

Where I_o is the absorbed intensity
g the effective thermal conduction to the environment.
ω the chopping frequency.
c the volume specific heat
d the element thickness.

Clearly from equation 1 for high sensitivity the numerator I_o should be as large as possible and systems have been designed using radiation concentrators, thermal collectors and antenna structures. Very effective highly absorbing blacks of exceedingly low thermal mass have also been designed.

To further improve responsivity the denominator must be kept very small. Thermal isolation (g), thermal mass (cd) and chopping frequency must be simultaneously addressed. Contrary to intuition excellent thermal isolation (very low g) is an absolute pre-requisite to permit ultra low thermal mass (very small detector thickness) and a reasonable chopping frequency, together with high sensitivity.

In isolated single (point) detectors, thermal isolation is easy as the element may be mounted on a polymer film drum head and connected by thin evaporated metal leads. For imaging systems isolation of the individual pixels is much more difficult and a major problem in most current hybrid systems.

(b) Many material systems change their properties rapidly with temperature and can be used as thermometric elements. Gases, as in the golay cell, one of the earliest high sensitivity thermal sensors, thermistors primarily simple NTC compositions [38], but high Tc superconductors have recently been applied [39], optical refractive index as in recent pyrooptic studies, but of most relevance to this paper pyroelectric and dielectric properties.

In the simple pyroelectric sensor, it is the differential change of remanent polarization with temperature $\partial P/\partial T$ which provides a current source

$$i = \frac{\partial P}{\partial T} \cdot \frac{\partial T}{\partial t} \quad (9)$$

where $\partial T/\partial t$ is the rate of change of temperature. $\partial P/\partial T$ is called the pyroelectric coefficient p.

To achieve a voltage output the pyroelectric must charge its self capacitance

$$\Delta V = \frac{\Delta Q}{C} = p_1 \; A\Delta T/C \quad (10)$$

for thin sheet detector polarized normal to the plane

$$C = \varepsilon A/t \quad (11)$$

$$\frac{\partial V}{\partial T} = \frac{p_1 \quad t}{\varepsilon} \qquad (12)$$

If the dielectric permittivity is a strong function of temperature as in ferroelectrics near the Curie transition additional response may be achieved by operating the material under bias and inducing an additional pyroelectric p_2 where

$$p_2 = \left(\frac{\partial \varepsilon}{\partial T}\right) E_b \qquad (13)$$

and

$$\frac{\partial V}{\partial T} = \frac{(P_1 + P_t) \quad t}{\varepsilon} = \left(\frac{\partial \varepsilon}{\partial T} \cdot V + t \frac{\partial P}{\partial T}\right) \varepsilon^{-1} \qquad (14)$$

It has become conventional to develop material figures of merit for pyroelectric and dielectric systems.

For current mode operation $F_i = P_T/\varepsilon C^1$ where C^1 is volume specific heat.

For voltage made operation $F_V = P_T/\varepsilon C^1$ and for high signal to noise.

$$F_d = Pt/C^1(\varepsilon \tan \delta)^{1/2} \qquad (15)$$

recognizing that the loss tangent δ is an effective noise source endemic to the dielectric.

An excellent summary of the figures of merit for a very wide range of pyroelectric and biased dielectric materials has been tabulated by Kulwicki et. al. [40] and is reproduced here as Tables II and III.

Point detectors are a mature technology used in a very wide range of consumer products, intruder alarms, proximity detectors, level sensors, etc. Frequently used materials include $LiTaO_3$, lead germanate (PGO) PLZT and the polymer PVDF. For very high sensitivity units such as horizon sensors in spacecraft the water soluble TGS family is still used and detectors which have sensitivities approaching within a factor of three of ultimate thermal noise limited performance have been produced.

In current imaging systems hybrid technology is used with the sensing plane reticulated into pixel elements that are solder bump bonded to the silicon MOS image processing electronics. Materials competition is between barium strontium titanate and lead scandium tantalate polished to thickness between 10 and 25 μ meters. Noise equivalent temperatures of 0.07°C in the object plane have been achieved with F1 optics. It is perhaps an interesting comment on the current social scene that the major commercial market for imagers in USA is in law enforcement, whilst in England it is to fire brigades for seeing victims in smoke filled rooms [41]!

Markets for current imagers are limited by the difficulty and high cost of the hybrid approach. With the advent of ferroelectric thin films there has been a strong drive to lower cost by using integrated approaches and depositing films on an air bridge thermal isolation system fabricated by MEMS technology. Initial results look most promising and it may be that by this approach thermal imaging may supplement night vision in automobiles and contribute to collision avoidance in fog.

TABLE II. Normal Pyroelectric ($T < T_c$)

Material	Bias kV/cm	T_c ^0C	c′ J/cm³K	p μC/cm²K	ε_r	tan δ	F_v cm²/C	F_d (cm³/J)$^{1/2}$	Reference
Single crystals									
TGS	0	49	2.3	0.028	38	0.01	3620	0.066	Herbert [16]
DTGS	0	60	2.4	0.055	43	0.02	6020	0.083	Whatmore (Felix) [25]
ATGSAs	0	51	2.5	0.07	32	0.008	9900	0.19	Bhalla [38]
$LiTaO_3$	0	665	3.2	0.18	43	0.003	1440	0.051	Byer [35]
$LiNbO_3$	0	1210	3.0	0.083	28	0.005	1140	0.025	Byer [35]
SBN 46/54	0	132	2.1	0.043	380	0.003	610	0.065	Byer [35]
PGO:Ba	0	70	2.0	0.032	81	0.001	2200	0.19	Whatmore (Watton) [39]
Ceramics									
PLZT 7/65/35	0	150	2.6	0.13	1900	0.015	300	0.032	Liu [41]
PLZT 8/65/35	0	105	2.6	0.18	4000	0.003	190	0.066	Deb [42]
PZNFTU	0	230	2.7	0.039	290	0.0031	570	0.052	Osbond [23]
PSZNFTU	0	170	2.7	0.049	400	0.0028	520	0.058	Osbond [23]
PGO	0	178	2.6	0.002	25	0.003	350	0.009	Kanduser [43]
Polymerics									
PVDF	0	none	2.4	0.0027	12	0.015	1040	0.009	Whatmore [39]
Thin films									
$PbTiO_3$ sol-gel	0	490	2.9	0.095	200	0.02	1870	0.056	Polla [28]
PLT 90/10 sputtered	0	330	3.2	0.065	200	0.006	1150	0.062	Takayama [44]
PCT 70/30 sputtered	0	270	3.3	0.05	390	0.015	440	0.021	Yamaka [45]
PZT 54/46 sol-gel	0	380	3.1	0.07	950	0.016	260	0.019	Ye [28]

Table III. Phase Transition Materials ($T \cong T_{t(c)}$)

Material	Bias kV/cm	T_t ^0C	c′ J/cm³K	p'_{max} μC/cm²K	ε_r	tan δ	F_v cm²/C	F_d (cm³/J)$^{1/2}$	Reference
Single crystals									
DTGFB	0	74	2.0	1.4	2400	0.02	3300	0.34	Dougherty [36]
KTN 67/33	2.5	4	3.7	8.0	25000	0.002	1000	0.46	Stafsudd [46]
BST 65/35	5	5	2.5	0.3	20000	0.007	170	0.034	Whatmore [39]
Ceramics									
BST 67/33	1	21	3.2	23.0	31000	0.028	2700	0.84	This work
BST 67/33	2	22	3.2	6.3	33000	0.021	670	0.25	This work
BST 67/33	6	24	3.2	0.70	8800	0.004	280	0.12	This work
BST 65/35	40	29	2.5	0.10	1200	0.0013	380	0.11	Whatmore [39]
PMN:La	90	40	3.0	0.085	1200	0.0008	800	0.10	Whatmore [39]
PScT	53	40	2.7	0.38	2900	0.0027	550	0.17	Shorrocks [47]
PZT 94/6*	0	50	2.6	0.37	300	0.02	5300	0.19	Lian [48]
PZNT 90/8/2*	0	30	2.6	0.185	290	0.019	2800	0.10	Takenaka [49]
PCT 70/30:PCoW 96/4	0	106	3.3	3.0	7400	0.037	1380	0.18	Deb [52]
PLZT 8/60/40 (*sic*)	0	142	3.7	3.2	16200	0.062	600	0.092	Deb [50]
PZN/BT/PT 80/10/10**	0	12	2.7	5.93	4670	0.01	5310	1.08	Deb [51]
PZN/BT/PT 80/10/10	0	85	2.7	2.9	18300	0.018	660	0.20	Deb [51]
Thin films									
PScT sputtered	40	40	2.7	0.52	6000	0.012	360	0.076	Watton [24]
PScT sol-gel	--	40	2.7	0.30	7000	0.002	180	0.10	Shorrocks [20]
PScT MOCVD	20	40	2.7	0.08	1000	0.002	330	0.07	Ainger [53]
KTN metalorganic	30	40	3.7	20.0	1200	0.01	50000	5.0	Schubring [54] "pumped"

* rhombohedral-rhombohedral phase transition
** rhombohedral-tetragonal phase transition

ELECTRO-OPTICS

In paraelectric and ferroelectric crystals the high intrinsic dielectric softness means that large internal (cooperative) ion motions can be induced by quite modest electric fields. Changes of mutual separation of the ions lead to strong distortions of the electronic charge distribution and hence strong changes in the optical refractive index and correspondingly large electrooptic effects. It is interesting to note that in these systems it is the electric field related constants which are exceptionally large and the polarization related constants quite normal.

For a single crystal the optical refraction behavior is defined by the optical indicatrix

$$\frac{x_1^2}{n_1^2} + \frac{x_2^2}{n_2^2} + \frac{x_3^2}{n_3^2} = 1 \qquad (16)$$

where $n_1 n_2 n_3$ are the principle refracture indices. If the optical impermeability is defined by

$$B_{11} = \frac{1}{n_1^2} \qquad B_{22} = \frac{1}{n_2^2} \qquad B_{33} = \frac{1}{n_3^2} \qquad (17)$$

Then the relationship may be expressed simply in tensor form

$$B_{ij} x_i x_j = 1 \qquad (18)$$

The perturbation to the impermeability ΔB_{ij} by electric field components $E_k E_l$ then takes the form

$$\Delta B_{ij} = f^X_{ijk} \; E_k + L^X_{ijkl} \; E_k E_l \qquad (19)$$

The f_{ijk} are the linear electrooptic tensor coefficients measured at constant stress (superscript X) and the L^{X}_{ijkl} are the quadratic electrooptic tensor coefficients measured again at constant stress.

For non zero f constants the crystal must be in a non centric piezoelectric group, and the f carry the same symmetry constraints as the piezoelectric d constants. The quadratic L constants have the same constraints as the electrostriction tensor and all transparent crystal have non zero quadratic constants.

Since the electric field produces elastic distortion, the f constant encompasses a stress/strain optic component in the form

$$\Delta B_{ikj} = e^{x}_{ijk} \ E_k + P^{E}_{ikjl} \ x_{kl} \qquad (20)$$

where p_{ikjl} are the stress optic tensor coefficients and x_{kl} the piezoelectrically induced strain components.

Awareness of the stress optic effects are of major importance as for a crystal under AC field at low frequency the system is free to deform and measurement give the constant (zero) stress f constant. However, as the crystal passes through elastic resonances it becomes successively more inertia clamped and the measurement drops back to the e constants. Because of the very high piezoelectric and electrostriction constants in ferroelectrics the change from f to e can be substantial.

As the dielectric constant goes through orders of magnitude changes with temperature in paraelectric and ferroelectric phases the constants f and M are similarly temperature dependent. In terms of the polarization components (equation 19) may be rewritten

$$\Delta B_{ij} = r_{ijk} \ P_k + M_{ijkl} \ P_k P_l \qquad (21)$$

The r and M constants are now similar to those in normal non ferroelectric oxides and only change slowly the temperature. For perovskite oxides, these constants are continuous through the ferroelectric phase transitions and optical changes between para and ferro phases can be calculated by simply inserting the appropriate non zero components of spontantous polarization P_kP_l into the equation for the prototypic form.

Polycrystal ceramic oxides are usually opaque to visible light due to strong scattering at grain boundaries and pores. In the lead lanthanum zirconate titanate (PLZT) ferroelectric and relaxor ferroelectric compositions it was found to be possible to hot press to near theoretical density and produce highly transparent ceramics [42]. With increasing lanthanum concentration the Curie range goes to lower temperature so that in a 65:35 Zr/Ti composition at 9 mole% La_2O_3 the system will not support a remanent polarization and the electrooptic effects are quadratic [43]. Between crossed polarizers thin sheet PZT is opaque, but can be biased under moderate field to highly transmitting. The switching speed is fast enough to provide protection against flash blindness and these materials are the basis for the Airforce Nuclear flash blindness helmets supplied to nuclear bomber pilots [44]. In commercial application however, the halfwave (transmitting) voltage is a little too high and the material too expensive to compete with ferroelectric liquid crystal systems.

For single crystals applications in simple electro-optics it is necessary to be able to grow large single crystals. Potassium dilhydrogen phosphate (KDP) has been widely used in its paraelectric phase above Tc for large area modulators. In guided wave structures lithium niobate is the crystal of choice. For the niobate crystals the refractive index may be enhanced by indiffusion of titanium so that guided wave structures can be fabricated on the crystal surface and a wide range of modulators, switches and other components are currently fabricated by this means.

In the tungsten bronze structure $Ba_{0.4}Sr_{0.6}Nb_2O_6$ (SBN) large optically perfect single crystals have now been grown. An advantage for SBN is that its stress optic coefficients are sufficiently large that guiding structures may be fabricated by local stress generated by capping films, preserving the high optical quality of the guiding path [45].

For lithium niobate, $BaTiO_3KNbO_3$ and SBN, high intensity laser light can promote carriers from traps in the band gap leading to a photo current. Because of the spontaneous polarization in the ferroelectric state the current flow is not symmetrical and re-trapping leads to an internal field which modulates the refractive index through the high electro-optic coefficients. A very wide range of research are presently exploring potential application for the strong photo refractive effects in these crystals [46].

Optical phase conjugation [47], holographic interferometry [48], image amplification at low light levels [49], optical beam fanning [50], moving object correlation [51], spacial soliton self trapping [52], and dynamic tuneable ferroelectric domain gratings for quasi phase matched harmonic generation [53] have all been generated at the research level.

For thin film systems optical applications have so far proved difficult as the quality and homogeneity of most present ferroelectric films is far from ideal. With the new capabilities which are being developed using Chemical Vapor Deposition (CVD) to produce truly epitaxial single crystal films it should be possible in future to exploit the capability that this will afford to produce excellent quality crystals with compositionally 'tuned' properties most suitable for specific optical applications.

REFERENCES

1. L.E. Cross, "Ferroelectric Ceramics: Tailoring Properties for Specific Applications," Ferroelectric Ceramics, Monte Verita, Edited by N. Setter, E. Colla, Birkhausen, Verlad, Basel (1993).

2. A.J. Bell, Multilayer Ceramic Processing, Ibid.

3. L.E. Cross and K.H. Hardtl, "Ferroelectrics," Kirk Othner Encyclopedia of Chemical Technology, John Wiley and Sons (1980).

4. K. Kinoshita and A. Yamaji, "Grain Size Effects on the Dielectric Properties in Barium Titanate Ceramics," J. Appl. Phys. **47**, 371 (1976).

5. L.E. Cross, "Relaxor Ferroelectrics," Ferroelectrics **76**, 241 (1987).

6. C. Randall, D. Barker, R. Whatmore, and P. Groves, "Short Range Order Phenomena in Lead Based Perovskites," Ferroelectrics **76**, 277 (1987).

7. P. Asadipour, "Polarization Mechanisms in Relaxor Ferroelectrics," MS Thesis, The Pennsylvania State University (May 1986).

8. D. Viehland, S.J. Jang, L.E. Cross, and M. Wuttig, "Freezing of the Polar Fluctuations in Lead Magnesium Niobate Relaxors," J. Appl. Phys. **68**, 2916 (1990).

9. D. Viehland, S.J. Jang, L.E. Cross, and M. Wuttig, "Local Polar Configurations in Lead Magnesium Niobate Relaxors, "J. Appl. Phys. **69**, 414 (1991).

10. S.W. Choi, T.R. Shrout, S.J. Jang, and A.S. Bhalla, "Dielectric and Pyroelectric Properties in the $Pb(Mg_{1/3}Nb_{2/3})O_3$:$PbTiO_3$ System," Ferroelectrics **100**, 29, 1989.

11. D. Viehland, M. Wuttig, and L.E. Cross, "The Glassy Behavior of Relaxor Ferroelectrics," Ferroelectrics **120**, 71 (1991).

12. J. Chen, "Electrical and Electromechanical Properties of Ferroelectric Thin Films for MEMS Applications," PhD Thesis, The Pennsylvania State University (August 1993).

13. R. Gerson and T.C. Marshall, "Dielectric Breakdown in Porous Ceramics," J. Appl. Phys. **30**, 1650 (1959).

14. W. Williams, III, "Investigation of Thin Film Ferroelectric Materials for Application in High Frequency Decoupling Capacitors," ISIF 1995, Colorado Springs, CO (March 20, 1995).

15. V. Chivukula, "Dielectric Properties of Ferroelectric Thin Films in the Frequency Range Hz to Ghz," ISIF 1995, Colorado Springs, CO (March 20, 1995).

16. L.E. Cross, "Dielectric Design in Ferroelectric Thin Films," Proceedings Electroceramics IV, Aachen Germany (September 5, 1994).

17. K.G. Brooks, J. Chen, K.R.Udayakumar, and L.E. Cross, " Electric Field Forced Switching in la Modified Lead Zirconate Stannate Titanate," J. Appl. Phys. **75** (3), 1699 (1994).

18. L.E. Cross, "Induced Strain Response in Ferroelectric Relaxor Ferroelectric on Phase Switching Films," Nato Advanced Research Workshop, Villa del Mar, Italy (June 20, 1994).

19. B. Jaffe, W.R. Cooke, Jr., and H. Jaffe, "Piezoelectric Ceramics," Academic Press, NY (1971)

20. E. Sawaguchi, "Ferroelectricity versus Antiferroelectricity in Solid Solutions of $PbZrO_3$ and $PbTiO_3$," J. Phys. Soc. Japan **8**, 515 (1953).

21. B. Jaffe, R.S. Roth, and S. Marzullo, "Properties of Piezoelectric Ceramics in the Solid Solution Lead Titanate - Lead Zirconate Lead Oxide Tinoxide and Lead Titanate Lead Hafnate," J. Res. National Bureau of Standards 55,2626 (1955).

22. W. Cao and L.E. Cross, "Distribution Function of Coexisting Phases in Complete Solid Solution System," J. Appl. Phys. **73**, 3250 (1993).

23. M.J. Haun, E. Furman, S.J. Jang, and L.E. Cross, "Thermodynamic Theory of the Lead Zirconate-Titanate Solid Solution System, Part 1: Phenomenology," Ferroelectrics **99**, 13 (1989).

24. M.J. Haun, E. Furman, H.A. McKinstry, and L.E. Cross, "Thermodynamic theory of the Lead Zirconate-Lead Titanate Solid Solution System: Part II-Tricritical Behavior," Ferroelectrics **99**, 27 (1989).

25. M.J. Haun, Z.Q. Zhuang, E. Furman, S.J. Jang, and L.E. Cross, "Thermodynamic Theory of the Lead Zirconate-Lead Titanate Solid Solution System: Part III- Curie Constant and Sixth Order Polarization Interaction Dielectric Stiffness Coefficients," Ferroelectrics **99**, 45 (1989).

26. M.J. Haun, E. Furman, T.R. Halemane, and L.E. Cross, "Thermodynamic Theory of the Lead Zirconate-Lead Titanate Solid Solution System: Part IV-Tilting of the Oxygen Octahedra," Ferroelectric **99**, 55 (1989).

27. M.J. Haun, E. Furman, S.J. Jang, and L.E. Cross, "Thermodynamic Theory of the Lead Zirconate-Lead Titanate Solid Solution System: Part V-Theoretical Calculations," Ferroelectrics **99**, 63 (1989).

28. K. Carl and K.H. Hardtl, "Stabilization of Domain in Acceptor Doped PZT Ceramics," Ferroelectrics **17**, 472 (1978).

29. D. Dederichs and G. Arlt, "Isothermal Aging in Iron Modified Lead Zirconate Titanate Ceramics," Ferroelectric **68**, 281 (1986).

30. R. Gerson and H. Jaffe, "Domain Processes in Lead Titanate Zirconate Ceramics," J. Phys. Chem. Solids **24**, 979 (1963).

31. R. Gerson, "Donor Doped Lead Zirconate:Titanate Ceramics," J. Appl. Phys. **31**, 188 (1960).

32. L.E. Cross, S.J. Jang, R.E. Newnham, S. Nomura, and K. Uchino, "Large Electrostriction Effects in Relaxor Ferroelectrics," Ferroelectrics **23**, 187 (1980).

33. Q. Zhang, W. Pan, A. Bhalla, and L.E. Cross, "Electrostrictive and Dielectric Response in Lead Magnesium Niobate Lead Titanate (PMN 0.1PT) and Lead Lanthanum Zirconate Titanate (PLZT 9.5/65/35) Under Variation of Temperature and Electric Field," J. Am. Ceram. Soc. **72**, 599 (1989).

34. R. Aldrich, ITEK Optics Division, Private Communication.

35. R.E. Newnham, D.P. Skinner, and L.E. Cross, "Connectivity and Piezoelectric/Pyroelectric Composites," Mat. Res. Bull. **13** (5), 525 (1978).

36. A.A. Shaulov, M.E. Rosar, W.A. Smith, and B.M. Singer, "Composite Ultrasonic Transducers" Proc. ISAF86, pp. 231, Lehigh University (1986).

37. W.A. Smith, "Composite Piezoelectric Materials for Medical Ultrasonic Imaging Transducers. A Review," Proc. ISAF86, pp. 249, Lehigh University (1986).

38. K.C. Liddiard, "Thin Film Resistance Bolometer IR Detectors," Infrared Physics **24**, 57 (1984).

39. M.C. Foote, B.R. Johnson, and B.D. Hunt, "Transition Edge $YB_2Cu_3O_7$ Microbolometers for Infrared Staring Arrays," Proc. SPIE (January 1994).

40. B.M. Kulwicki, A. Amin, H. Beratan, and C.M. Hanson, "Pyroelectric Imaging," Proc. ISAF92, Greenville, South Carolina (August 30, 1992).

41. R. Watton, "Ferroelectric Materials and IR Bolometric Arrays: From Hybrid Array Towards Integration," Proc. ISIF5, Colorado Springs, Colorado (April 1993).

42. G.H. Haertling, "Improved Hot-Pressed Electro-Optic Ceramics in the $(PbLa)(ZrTi)O_3$ System," J. Am. Ceram. Soc. **54**, 303 (1971).

43. J.T. Cutchen, J.O. Harris, and G.R. Laguna, "Electro-optic Devices Using Quadratic PLZTs," Sandia Lab Technical Report SLA-73-0777 (1973).

44. J.T. Cutchen, "PLZT Thermal/Flash Goggles Concepts and Constraints," Ferroelectrics **27**, 173 (1980).

45. H. Taylor, Texas A&M, Private Communication.

46. A.M. Glass, "Optical Materials," Science **235**, 1003 (1987).

47. G. Salamo, M.J. Miller, W.W. Clark, G.L. Wood, and E.J. Sharp, "Strontium Barium Niobate as a Self-Pumped Phase Conjugator," Optical Comm. **39**, 417 (1986).

48. N. Kukhtarev, Bo Su Chen, P. Venkateswarlu, G. Salamo, and M. Klein, "Reflection Holographic Gratings in [111] Cut $Bi_{12}TiO_{20}$ Crystal for Real Time Interferometry," Optical Comm. **104**, 23 (1993).

49. R.R. Neurgaonkar, W.K. Cory, J.R. Oliver, E.J. Sharp, G.L. Wood, and G.J. Salamo, "Growth and Optical Properties of Ferroelectric Tungsten Bronze Crystals," Ferroelectrics **142**, 167 (1993).

50. S.R. Montgomery, M.P. Gallagher, G.J. Salamo, E.J. Sharp, G.L. Wood, and R.R. Neurgaonkar, "Cooperative Photorefractive Beam Fanning in $BaSrKNaNb_5O_{15}$," J. Opt. Soc. Am. B **11**, 1694 (1994).

51. E.J. Sharp, G.L. Wood, G.J. Salamo, R.J. Anderson, Jan M. Yarrison-Rice, and R.R. Neurgaonkar, "Photorefractive Image Processing Using Mutually Pumped Phase Conjugators," SPIE Optical Pattern Recognition **2237**, 347 (1994).

52. M. Segev, B. Crosignani, A. Yariv, and B. Fischer, " Spatial Solitons in Photorefractive Media," Phys. Rev. Letters **68**, 923 (1992).

53. A.S. Kerwitsch, M. Segev, A.Yariv, G. Salamo, T.W. Towe, E.J. Sharp, and R.R. Neurgaonkar, "Tuneable Quasi Phase Matching Using Dynamic Ferroelectric Domain Gratings Induced by Photorefractive Space Change Fields," J. Appl. Phys. Letters **64**, 3068 (1994).

POROUS SILICA : A VERSATILE MATERIAL FOR MICROELECTRONIC APPLICATIONS

Rosario A. Gerhardt
School of Materials Science and Engineering
The Georgia Institute of Technology
Atlanta, GA 30332-0245

ABSTRACT

The potential applications of porous silica in microelectronic devices are varied. Some of the most promising include their use as interlevel dielectrics, chemical sensors, anti-reflective coatings and luminescent sources in integrated circuits, devices and multichip modules. While porous silica may be fabricated using a variety of processing methods, a homogeneous and reproducible porous microstructure is difficult to achieve. This paper discusses using colloidal processing and anodization of Si wafers as the two preferred methods because of their ease of fabrication and low cost. Some relevant property data is reported.

INTRODUCTION

Silica is ubiquitous in the semiconductor industry because it is the native oxide of the most important and widely used semiconductor silicon. This paper concerns itself with the potential applications of *porous* silica in microelectronic devices. Using porous silica as an interlevel dielectric has been under consideration for nearly 10 years [1-2] but its use in microelectronics has been limited due to difficulties in obtaining a homogeneous and reproducible pore microstructure and the inherent moisture adsorption the porosity entails. However, this material has the potential of serving a dual purpose, one as a very low permittivity dielectric (as low as 1.6[3]) and also as a built-in humidity sensor[4]. In addition, there are some additional applications such as using porous silica as an anti-reflective coating[5] and even as a visible luminescent source[6].

Porous silica thin films may be prepared via a variety of methods. These

To the extent authorized under the laws of the United States of America, all copyright interests in this publication are the property of The American Ceramic Society. Any duplication, reproduction, or republication of this publication or any part thereof, without the express written consent of The American Ceramic Society or fee paid to the Copyright Clearance Center, is prohibited.

include rf sputtering, e-beam evaporation, PECVD, CVD, sol-gel processing and anodization of Si wafers. Achieving controlled porosity by the first four methods is extremely difficult and costly therefore this work has been limited to the last two methods.

SOL-GEL PROCESSING

The main advantages of sol-gel processing are its low cost and compatibility with standard integrated circuit processing. Prior to gelation, the precursor materials may be dipped and/or spun in the same manner as polymeric resists and dielectrics. In addition, porous silica is more chemically compatible, has a better thermal expansion match and better thermal conductivity than conventionally used polyimides[7]. Many reports[1-3,7-11] have shown that porous silica can achieve as low or lower dielectric constants than the lowest permittivity polymers but in most cases moisture adsorption problems have led to short circuiting or at least a high dissipation factor[3]. These problems have kept porous silica from being used as insulating layers in hybrid circuits. Recent work[11] on xerogels indicates that porous silica thin films can be made hydrophobic. It is precisely this ability to be hydrophobic or hydrophilic that may be used to advantage in order to obtain insulating layers and humidity sensing layers in the same device. One can envision a sequence of masking, coating, leaching, treating, re-masking, etc in order to obtain the desired performance.

One of the challenges in sol-gel processing of porous silica thin films is to dry films of the desired thickness crack-free. For alkoxide type xerogels and aerogels, hypercritical drying is necessary[12]. This is not the case for colloidal processing since the average pore sizes obtained with this method are larger. For the last ten years, the present author has conducted research on the processing[3, 13-14], microstructural characterization[15-19] and dielectric properties[3,8,20-23] of porous silica prepared via a colloidal processing method originally invented by Shoup and Wein[24]. With this method, one can make gels with a volume porosity as high as 85%. Most of our work has been done on bulk gels but we have recently been successful at making nearly crack-free porous silica thin films of 2-12 microns in thickness[3,14]. Our results have shown that the gels are homogeneous on a scale smaller than 1 μm[15-17] and that the dielectric constant can be as low as 1.6 in bulk[8] as well as in thin film form[3]. At the same time, we have shown[21-23] that modifying the pore surface can substantially enhance the humidity sensitivity of these materials[22].

The method used consists of mixing known amounts of colloidal silica with potassium silicate. The mixture is then hydrolyzed with formamide and water. Gels are labeled according to the ratio of colloidal silica to potassium silicate. Figure 1 depicts the microstructure of an ultramicrotomed thin section of a 20:80 porous silica gel. The dark regions represent the skeletal silica and the white regions represent the pore space. As can be seen, the microstructure is very homogeneous on a scale of 0.5 microns or less. Figure 2 displays the surface of a thin film at a much lower magnification in order to demonstrate that films are crack free upon drying. Unfortunately, leaching to remove the alkali impurities introduced during processing does introduce some cracking but not enough to create short circuiting[14]. Figure 3 demonstrates that a dielectric constant of 1.6 can be achieved in the MHz range. The dielectric loss is currently too high (3%) but we are currently developing the procedure for making the films hydrophobic. We have found that the dielectric properties of untreated films deteriorate rapidly when exposed to higher than 55% RH but may be lowered to their original value after an appropriate heat treatment without physical damage.

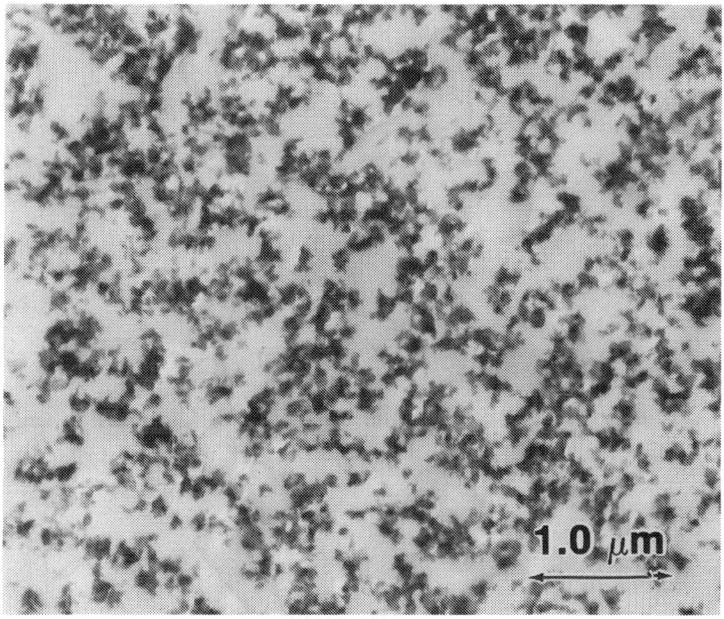

Figure 1. Ultramicrotomed thin section of a 20:80 colloidal silica: potassium silicate gel showing homogeneity of the pore space.

Figure 2. As dried porous silica thin film indicating absence of cracks and inhomogeneities over large areas.

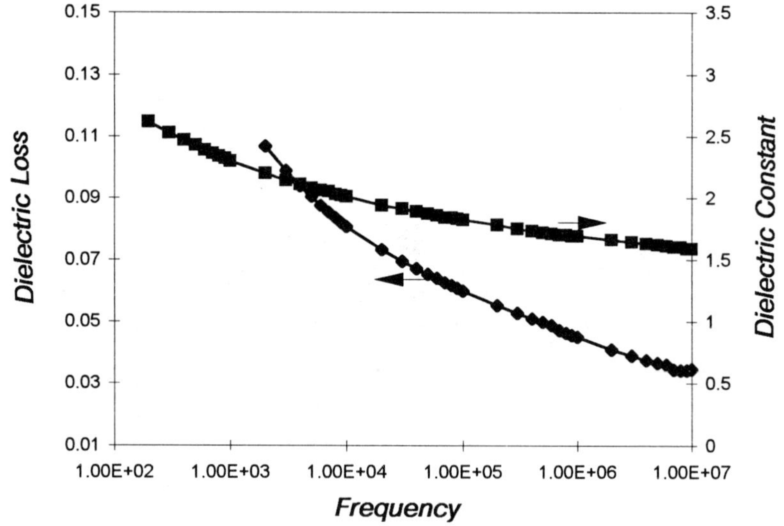

Figure 3. Dielectric constant and dielectric loss versus frequency for a 20:80 porous silica thin film.

As indicated earlier, surface doping porous silica prepared by the method described above dramatically enhances its humidity sensitivity[20-23]. The humidity sensitivity may be detected as an increase in the dielectric constant at lower frequencies, an increase in the loss or a decrease in the impedance[23]. Figures 4 and 5 show the effect of increasing humidity on a porous silica gel surface doped with 0.5 N K. Figures 4 and 5 clearly demonstrate that the dielectric constant increases and that the impedance decreases drastically as the humidity is increased from 39% to 90% RH. This impedance data may also be converted to conductivity by taking the real axis intercept on Figure 5 and multiplying it times the geometrical factors. The conductivity versus relative humidity obtained in such a manner is displayed in Figure 6 for porous silica gels that had been surface doped with 2 N solutions of various alkali salts[22]. The degree to which the conductivity increases can be tailor made to be a continuous increase or a step increase. Doping solution concentration and type of alkali ion are two of the possible ways to modify the level of sensitivity.

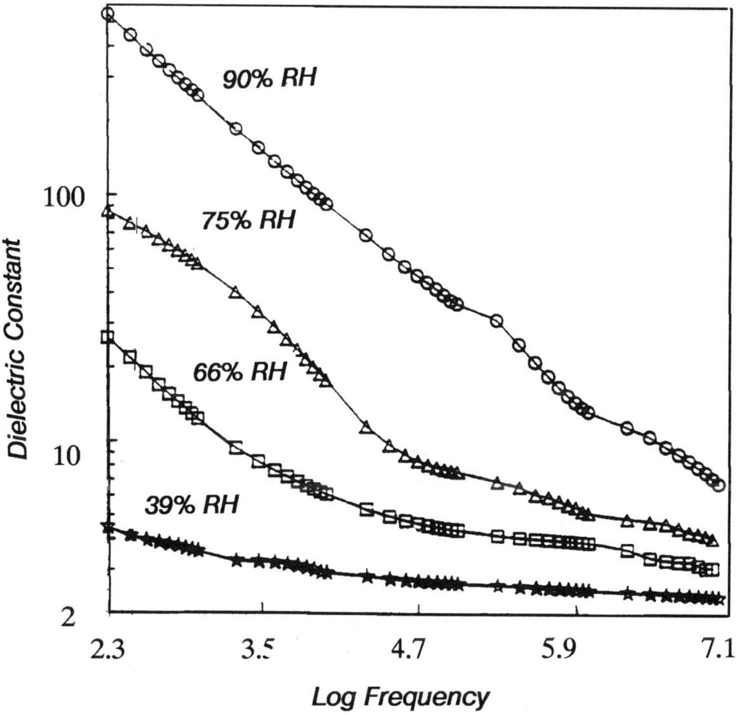

Figure 4. Effect of humidity on dielectric constant versus log frequency for a 15:85 porous silica gel surface doped with 0.5 N K.

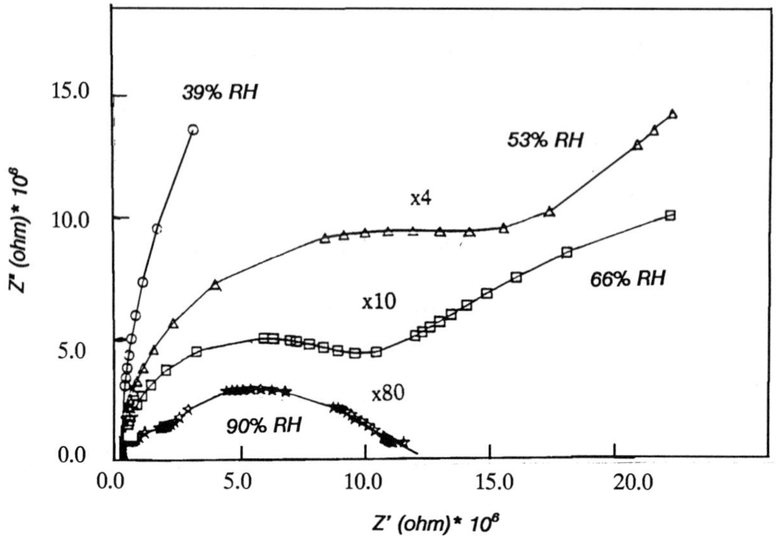

Figure 5. Effect of humidity on the complex impedance plot of a 15:85 porous silica gel surface doped with 0.5 N K.

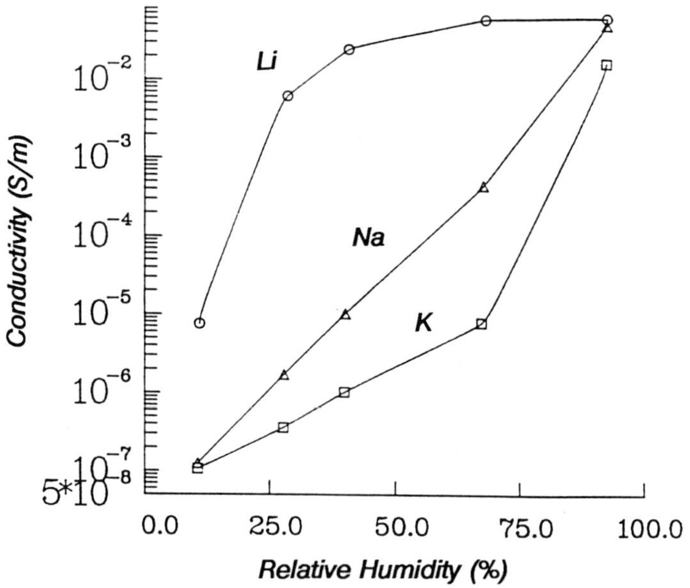

Figure 6. Conductivity versus relative humidity for three different 15:85 porous silica samples surfaced doped with 2 N of Li, Na and K respectively.

ANODIZATION OF Si WAFERS

The main advantage of making porous silica thin films via anodization of Si wafers is that this method has the possibility of being incorporated into integrated circuit designs more readily. The main challenge is to control the processing parameters so as to achieve the porosity content and properties desired. Since the discovery of visible luminescence in porous silicon[25], a lot of research has been conducted to achieve better luminescent properties. Not much attention has been paid to the actual porosity obtained or to the fact that the porous Si layers are actually porous silica since the high surface area created after anodization leads to rapid oxidation. Fauchet and co-workers[26] have done a careful job of documenting the effect of volume porosity on the photoluminescent properties. In the 1970s, some researchers[27] considered using porous Si(a misnomer) as an interlevel dielectric but the idea has not yet been implemented possibly due to difficulties in reproducibility and lack of mechanical robustness. Moisture adsorption problems could also have prevented it from being used in this application. However, just as in the case of porous silica thin films fabricated by sol-gel processing, moisture pick up could be used as an advantage for making built-in humidity sensors in a microelectronic device designed to perform a different function.

It has been reported in the literature that anodization factors such as current density, time and solution concentration have an influence on the photoluminescence obtained[28], hence the porosity can be expected to be affected as well. Other researchers have reported that the wafer dopant concentration also has an effect but little has been said about the effect of the crystallographic orientation and the surface condition of the wafers prior to anodization. Figure 7, which is an anodized polysilicon wafer, demonstrates that the amount and orientation of the porosity obtained is a function of the crystallographic orientation and its surface condition. It should be noted that the grain boundary regions are preferentially etched but that each crystallite orientation develops a different pattern. These differences have also been observed on single wafers of different orientation. Figure 8 shows a high magnification of a <111> wafer anodized at 100 mA/cm^2 for 10 minutes which demonstrates that a high level of uniformity is achievable on a scale finer than a micron. Having the ability to vary the scale of the porosity and controlling the areas to be anodized either by photomasking or inducing faster etching should be very useful for the application of porous silica in microelectronic devices[29,30]. Specific applications may include insulating regions, conducting conduits, built-in chemical sensors, anti-reflective coatings

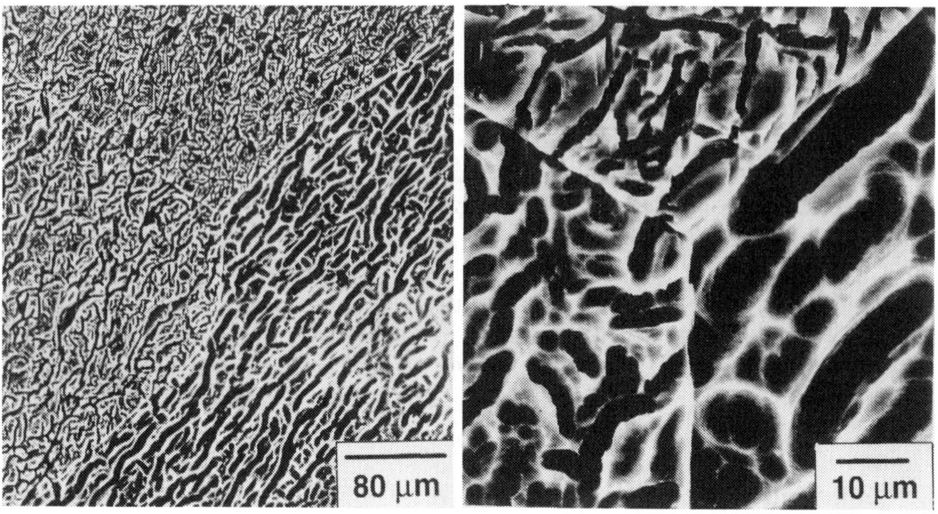

Figure 7. SEM micrograph of an anodized polysilicon wafer. Note that each crystallite has its own preferential etching pattern and that the grain boundary regions have etched away more quickly than the bulk grains.

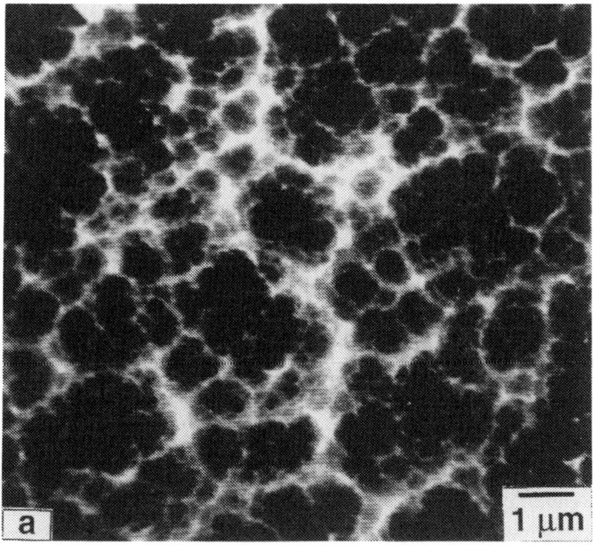

Figure 8. SEM micrograph of a <111> Si wafer anodized at 100 mA/cm^2 for 10 minutes. Notice homogeneous pore distribution at scale <1 μm.

and possible electroluminescent pixels. It is anticipated that the level of porosity and distribution needed may not be the same for all of these different applications.

In order to illustrate the versatility of this material, photoluminescence and reflectance spectra have been taken. Figures 9 and 10 portray photoluminescence versus wavelength data for anodized wafers of different crystallographic orientation and surface condition respectively. Figure 9 indicates that under the same anodization conditions, the crystallographic direction of the Si wafer has an effect on the magnitude and the position of the photoluminescence peak obtained. The influence of the surface condition prior to anodization is clearly demonstrated in Figure 10. Even though this is not surprising, the author is not aware of any documented evidence that can explain why the photoluminescent peaks shift in position with different starting surface conditions. A similar dependence on the surface condition of the photoluminescence spectra of ZnSe single crystals has also been observed[31]. It should be added that shifts from the red to the blue portion of the spectrum have been ascribed to chemical differences[32].

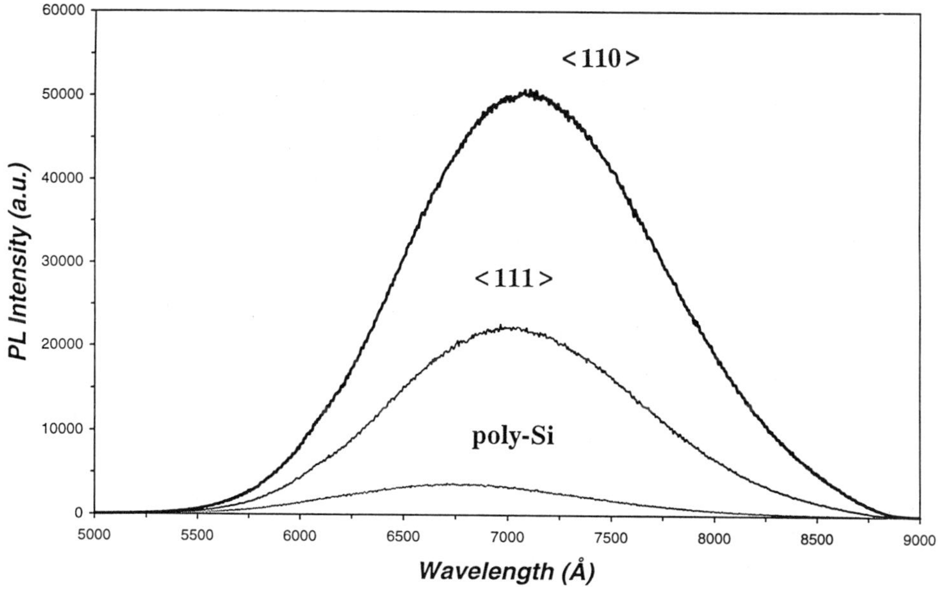

Figure 9. Photoluminescence intensity versus wavelength for single crystal wafers of different crystallographic orientation as compared to a polysilicon wafer anodized under the same conditions.

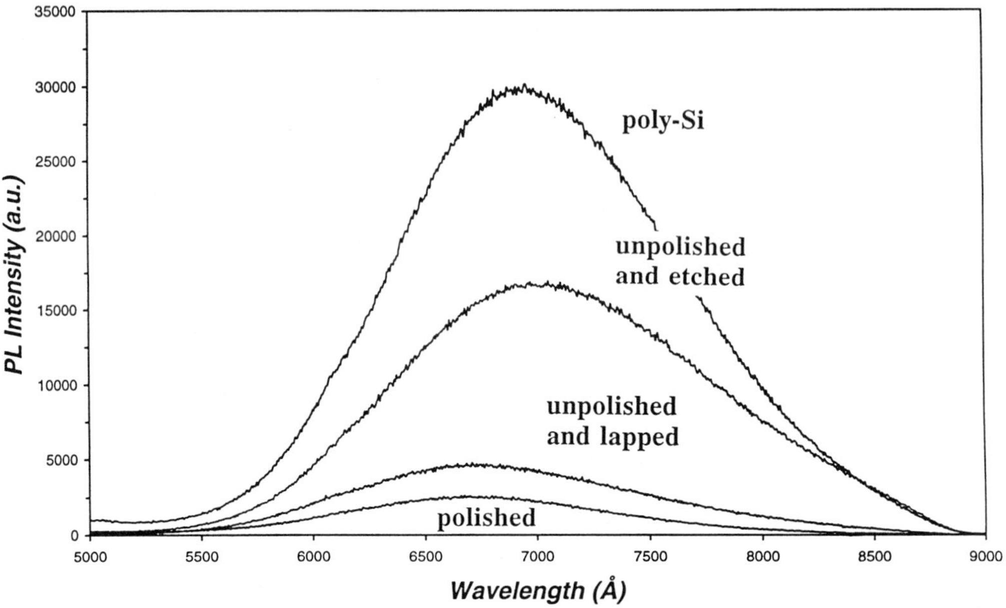

Figure 10. Photoluminescence spectra versus wavelength for <100> single crystal wafers anodized at 50 mA/cm^2 for 20 minutes after different surface treatments.

Reflectance measurements are of interest because the presence of porosity should lead to lower reflectance values. If successful, porous Si could be used in place of anti-reflective coatings in solar cells[33]. Figure 11 depicts the reflectance spectra for <100> oriented wafers that were anodized under different conditions. It can be seen that the lowest reflectance is obtained at 60mA/cm^2 for 30 minutes. In Figure 12, all samples were anodized under the same conditions of 50mA/cm^2 for 30 minutes but the surface condition was varied. Reflectance values are seen to decrease from approximately 40% for the polished wafer to 20% for one that was anodized after etching. Further reductions in reflectance are being explored.

CONCLUDING REMARKS

The above results suggest that the amount and distribution of the porosity have a large effect on the resultant properties examined. This presents both a challenge as well as the ability to tailor make porous silica films with specific properties. The two methods presented here are the most economical of all the methods available for the fabrication of porous silica and thus should be continued to be explored. Further work is needed in order to assess the best conditions for fabrication of porous silica for the various

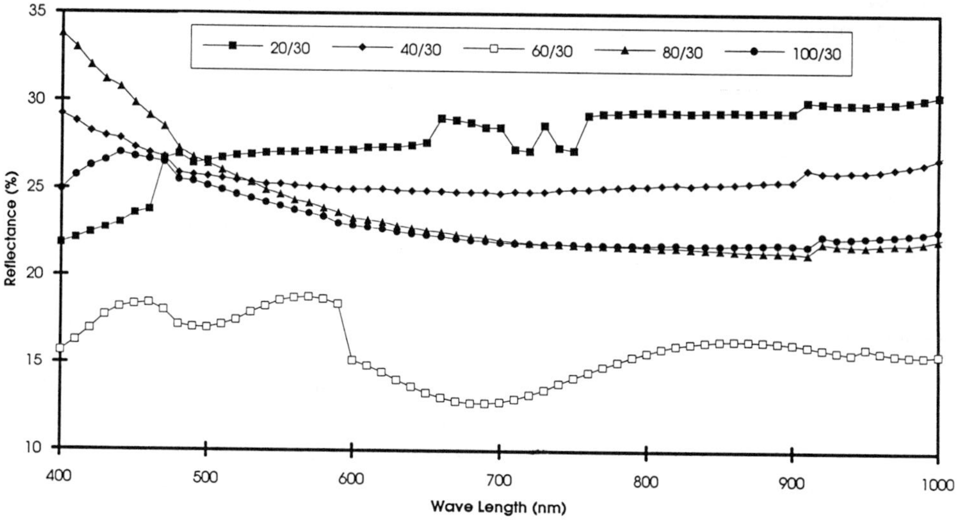

Figure 11. Reflectance versus wavelength for single crystal <100> wafers anodized at different conditions.

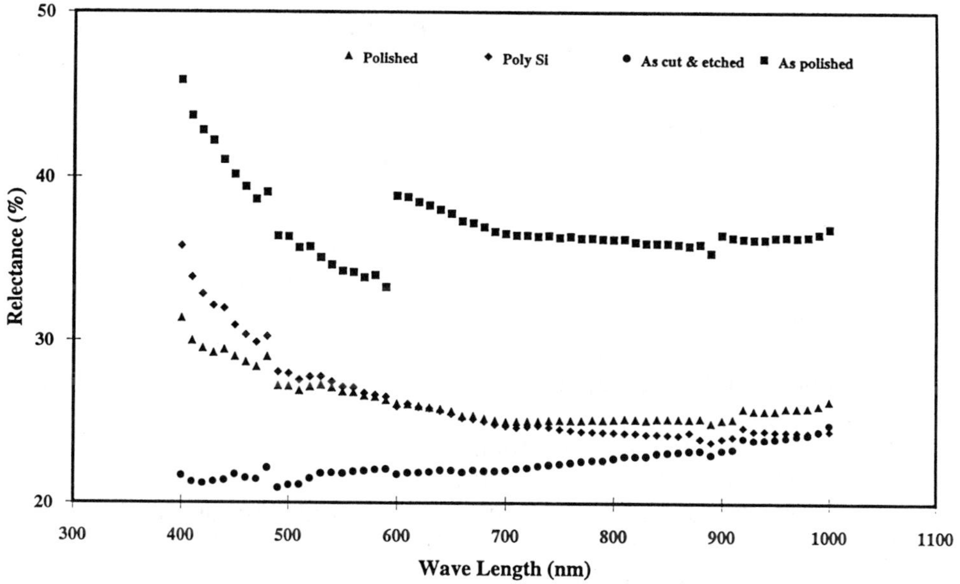

Figure 12. Effect of surface condition on the reflectance spectra of <100> single crystal wafers anodized at 50 mA/cm^2 for 30 minutes.

microelectronic applications mentioned. The biggest challenge for the implementation of porous silica as an interlevel dielectric will be to achieve and maintain hydrophobicity of the films without compromising their mechanical integrity and/or dielectric properties. This is a worthy goal in view of the fact that the semiconductor industry expects an interlevel dielectric with permittivity less than 2 by the year 2004[34]. For humidity sensors, the issues of aging and drift will be big considerations. Applications of porous silica as anti-reflective coatings and electroluminescent devices prepared by sol-gel processing and/or anodization of Si will need further optimization.

Acknowledgements

This work was funded in part by the Georgia Tech Research Corporation and the National Science Foundation under grant DMR-9311765. The author would like to acknowledge the contributions of M. Sankararaman, J.R. Kokan and G. Zhang and useful discussions with P.A. Kohl, A. Rohatgi and C. Summers.

REFERENCES

1. A. Das, R. Messier, T.R. Gururaja and L.E. Cross, "Low Permittivity SiO_2/Void Nanocomposite Films," in Electronic Packaging Materials Science II, *Mat.Res.Soc.Symp.Proc. 72*, 27-33 (1986).
2. U. Mohideen, T.R. Gururaja, L.E. Cross and R. Roy, "Ultra-Low Dielectric Constant Porous Silica Thick Films for High Speed IC Packaging," IEEE Trans. Comp. Hybrids and Manuf. Tech., vol. II[1] 159-162 (1988).
3. R.A. Gerhardt, J.R. Kokan and P.A. Kohl, "Low Permittivity Porous Silica Thin Films for MCM-C/D Applications," in Advances in Ceramics Multi-Chip Modules (MCM) and High Performance Electronic Materials, *NATO Workshop Proc.*, 209-219(1994).
4. J.R. Kokan and R.A. Gerhardt, "Sol-Gel Processing of Porous Silica Thin Films for Humidity Sensing Applications," unpublished results.
5. A. Rohatgi, private communication.
6. L. Tsybeskov, S.P. Duttagupta and P.M. Fauchet, "Photoluminescence and Electroluminescence in Partially Oxidized Porous Silicon," in Microcrystalline and Nanocrystalline Semiconductors, ed. R.W. Collins, C. C. Tsai, M. Hirose, F. Koch and L. Brus, *Mat.Res.Soc.Symp.Proc. 358*, 683-688 (1995).
7. R.A. Gerhardt, "Composites for Electronic Substrate Applications," in Electronic Packaging Materials Science III, ed. R. Jaccodine, K.A.

Jackson and R. C. Sundahl, *Mat.Res.Soc.Symp.Proc.108,* 101-106 (1988).
8. W. Cao, R. Gerhardt and J. B. Wachtman, Jr., "Low Permittivity Porous Silica by a Colloidal Processing Method," in Ceramic Substrates and Packages for Electronic Applications, ed. M. Yan, K. Niwa, H. O'Bryan, Jr. and W.S. Young, *Advances in Ceramics, Vol.26,* 409-418 (1989).
9. S.J. Stein, R. Wahlers, C. Huang, T. Grunstein and G. Sykora, "VLSI Packaging Materials Using Fireable Metallic and Dielectric Films," *Proc. International Microelectronics Conference,* SHM-Japan, pp. 1-8 (1990).
10. L.W. Hubresh, L.E. Keene and V.R. Latorre, "Dielectric Properties of aerogels," *J. Mat. Res. 8[7]* 1736-1741 (1993).
11. D.Smith, J. Anderson, C.C. Cho and B.E. Gnade, "Preparation of Low-Density Xerogels at Ambient Pressure for Low K Dielectrics," in Advances in Porous Materials, ed. S. Komarneni, D.M. Smith and J.S. Beck, *Mat.Res.Soc.Symp.Proc. 371*, 261-266 (1995)
12. C.J. Brinker and G.W. Scherer, Sol-Gel Science, Academic Press, 1990.
13. W. Cao, R. Gerhardt and J.B. Wachtman, "Preparation and Sintering of Colloidal Silica:Alkali Silicate Gels," *J.Am.Ceram.Soc. 71[12],* 1108-1113 (1988).
14. J.R. Kokan, F. Tadayon, R.A. Gerhardt and P.A. Kohl, "Sol-Gel Processing of Porous Silica Thin Films for Interlevel Dielectrics in Integrated Circuits," *IEEE Proc. 44th Annual Electron Components and Technology Conference*, pp. 704-711 (1994).
15. H.M. Kerch and R. Gerhardt, "Quantitative Electron Microscopic Investigation of the Pore Structure in 10:90 Colloidal Silica-Potassium Silicate Sol-Gels," *J.Am.Ceram.Soc. 73[8]*, 2228-2237 (1990).
16. H.M. Kerch, F. Cosandey and R. Gerhardt, "Imaging of Fine Porosity in Colloidal Silica: Alkali Silicate Gels by a Defocus Contrast Technique," *J. Non-Cryst. Solids 152* , 18-31 (1993).
17. S. Bhattacharja, F. D'Orazio, J. C. Tarczon, W. P. Halperin and R. Gerhardt, "Internal Structure of Porous Silica: A Model System for Characterization by Nuclear Magnetic Resonance," *J. Am. Ceram. Soc. 72 [11]* 2126-2130 (1989).
18. G.G. Long, S. Krueger, R.A. Gerhardt and R.A. Page, "Small Angle Neutron Scattering Characterization of Processing/Microstructure Relationships in the Sintering of Crystalline and Glassy Ceramics," *J. Mat. Res. 6 [12]* 2706-2715 (1991).
19. G.G. Long, S. Krueger, P.R. Jemian, D.R. Black, H.A. Burdette, J.P. Cline and R.A. Gerhardt, "Small Angle Scattering Determination of the Microstructure and Morphology of Porous Silica Precursors," *J. Appl. Cryst. 23*, 535-544 (1990).
20. W. Cao, R. Gerhardt and J.B. Wachtman, Jr., "Dielectric Relaxation of

Water Adsorbed on Monolithic Porous Silica Gels," in Ceramic Dielectrics, ed. H.C. Ling and M.F. Yan, *Ceram. Trans. vol. 8:* 175-184, 1990.
21. W. Cao, R. Gerhardt and J.B. Wachtman, Jr, "Effect of Sodium Ions on the Dielectric Conductivity of Porous Silica in Humid Environments," in Physical Phenomena in Granular Materials, *Mat.Res.Soc.Symp.Proc. 195,* 471-476 (1990).
22. R. Gerhardt, G. Zhang and W. Cao, "Analysis of Impedance Spectra for Dried Porous Silica Gels Exposed to Various Humidities", submitted to *Electrochimica Acta.*
23. R. Gerhardt, "Dielectric and Impedance Spectroscopy Revisited: Distinguishing Localized Relaxation from Long Range Conductivity," *J. Phys. Chem. Solids, 55[12]* 1491-1506 (1994).
24. R.D. Shoup and S. Wein, "Low Temperature Production of High Purity Fused Silica," *US Patent No. 4059658,* Nov. 22, 1977.
25. L.T. Canham, *Appl. Phys. Lett. 57*, 1046-1050 (1990).
26. L. Tsybeskov and P.M. Fauchet, *Appl.Phys.Lett. 64*[15]1983-1985(1994).
27. Y. Watanabe, Y. Arita, T. Yokoyama and Y. Igarashi, J. Electrochem.Soc. 122[10] 1351-1359(1975).
28. Various articles in Microcrystalline Semiconductors: Materials Science & Devices, ed. P.M. Fauchet, C.C. Tsai, L.T. Canham, I. Shimizu and Y. Aoyagi, *Mat.Res.Soc.Symp.Proc. 283* (1993).
29. C.R. Martin, "Nanomaterials: A Membrane-Based Synthetic Approach," *Science 266,* 1961-1966 (1994).
30. S.P. Duttagupta, P.M. Fauchet, C. Peng, S.K. Kurinec, K. Hirschman and T.N. Blanton, "Micron Size and Submicron size Light-Emitting Porous Silicon Structures," in Microcrystalline Semiconductors: Materials Science & Devices, ed. R.W. Collins, C.C. Tsai, M. Hirose, F. Koch and L. Brus, *Mat.Res.Soc.Symp.Proc. 358,* 647-652 (1995).
31. M.P. Volz, C.H. Su and R. Gerhardt, unpublished results.
32. L. Tsybeskov, Ju V. Vandyshev and P.M. Fauchet, "Blue emission in porous silicon: Oxygen-related photoluminescence," *Phys. Rev. B 49[11]* 7821-7824 (1994).
33. G. Sun, Y. Li, Y. Lu, B. Khan and G.S. Tompa, "Investigation of Efficiency Improvement on Silicon Solar Cells due to Porous Layers," in Microcrystalline Semiconductors: Materials Science & Devices, ed. R.W. Collins, C.C. Tsai, M. Hirose, F. Koch and L. Brus, *Mat.Res.Soc.Symp.Proc. 358,* 593-598 (1995).
34. R. K. Laxman, "Low E Dielectrics: CVD Fluorinated Silicon Dioxides," *Semiconductor International,*71-74, May 1995.

RAINBOWS AND FERROFILMS - SMART MATERIALS FOR HYBRID MICROELECTRONICS

Gene H. Haertling
Gilbert C. Robinson Department of Ceramic Engineering
Clemson University, Clemson, SC 29634-0907

ABSTRACT

This review paper describes the materials, processing, properties and applications of the newly developed ultra-high displacement Rainbows and thick/thin ferroelectric Ferrofilms. Their applicability to hybrid and fully integrated microelectronics is discussed in regard to each of these areas of concern.

INTRODUCTION

Current trends have shown that industrial and commercial hybrid microelectronic components designed for the automobile, home, office and factory are becoming an increasingly more important segment of present-day automation. The materials for such applications are required to be more sophisticated in that they must be able to perform more than one function (e.g., actuation and sensing) during operation or provide a unique combination of highly specialized properties. These materials are now commonly known as smart or intelligent materials and are exemplified by such general groups as ferroelectrics, piezoelectrics, pyroelectrics, electrooptics, electrostrictive materials and composites.

Key factors in the application of these materials to hybrid circuits and microelectronics are (1) their ability to be scaled down in size (i.e.,miniaturized or fully integrated) without loss of bulk properties, (2) their capability of achieving the proper form factor for the substrate and (3) their processing compatibility with other components on the substrate. Recent research and development work has shown that significant progress has been made in the last several years in each of these areas.[1-5] A new processing technique has recently led to the development of ultra-high displacement

To the extent authorized under the laws of the United States of America, all copyright interests in this publication are the property of The American Ceramic Society. Any duplication, reproduction, or republication of this publication or any part thereof, without the express written consent of The American Ceramic Society or fee paid to the Copyright Clearance Center, is prohibited.

ceramic actuators which have been coined as RAINBOWS (Reduced And INternally Biased Oxide Wafers), an acronym for the chemical reduction process used to transform ordinary planar piezoelectric and electrostrictive wafers into domed, high displacement, two-dimensional bending actuators of moderate load-bearing capability.[6-7] Rainbow devices such as speakers, pumps, switches, deflectors and linear actuators have been made in sizes as large as 10 cm. in diameter and as small as 2 mm. in diameter or length.

An assessment of the present-day ceramic actuator technologies for bulk materials is given in Table 1. As seen from the table, a variety of direct extensional configurations, composite flextensional structures and bending mode devices are used to achieve an electromechanical output. Trade-offs between stress generating/loading capability and displacement are commonly made when designing for particular applications. Maximum displacement can be seen to be achieved with composite or bender structures; however, this is usually accomplished at the expense of less load bearing capability, greater complexity and higher cost. The recent introduction of the Rainbow bender to this family of actuators has served to either extend the stress capability of the bender technologies without sacrificing displacement or extending the displacements achievable with equal load-bearing capacity. Additionally, because of their unique dome or saddle-type structure, small discrete Rainbow elements for hybrid circuits can be fabricated from larger, bulk processed wafers.

For the fabrication and integration of actuator/sensors and other devices at the microelectronic (micron) level, one must consider different technologies than those previously mentioned. Among these techniques are vacuum deposition, spinning, dipping, chemical vapor deposition and laser ablation; however, those which have been reported to successfully produce both thin (0.02 - 5 um in this paper) and thick (5 - 30 um in this paper) films of the above mentioned materials are considerably fewer in number, and the most promising of these, in the near term, is dipping. Using an automated dipping apparatus, Li, et.al.,[8] were successful in fabricating thin and thick films of PLZT ferroelectric and electrooptic compositions on Ag, Si, sapphire and glass substrates. Films as thick as 15 microns have yielded properties quite similar to those of the bulk material and show excellent promise for future devices.

It is the object of this review paper to describe the materials, processing and properties of two types of recently developed smart materials; i.e., Rainbows for discrete and hybrid structures and thick/thin ferroelectric films (Ferrofilms) for hybrid and integrated structures. Examples of typical applications are discussed.

Table 1. Present-Day Ceramic Actuator Technologies

Type	Configuration	Max. Stress* (MPa)	Actuator Movement (w/Voltage)	Actuator Type (P or E)	Actuator Displacement (%)+
Monolithic (d_{33} mode)	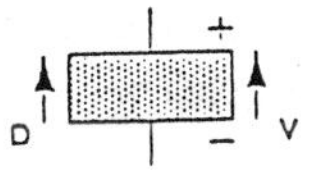	40	Expansion	P	0.40
Monolithic (d_{31} mode)	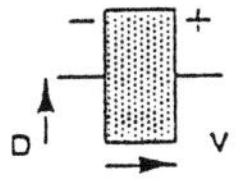	40	Contraction	P	-0.15
Monolithic (s_{11} mode)	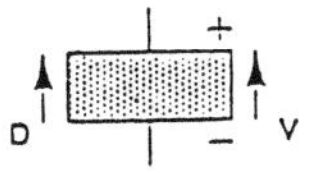	40	Expansion	E	0.24
Monolithic (s_{12} mode)	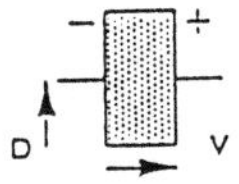	40	Contraction	E	-0.08
Composite Structure (d_{33} mode) (flexten.)	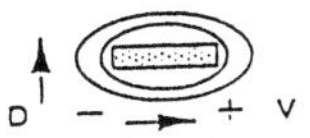	10	Contraction	P	-1.0
Composite Structure (d_{33}/d_{31}) (Moonie)	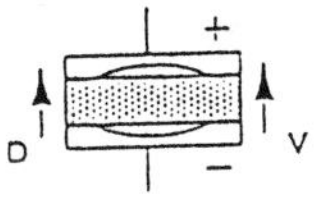	0.028	Expansion	P/E	1.3
Unimorph (bender)	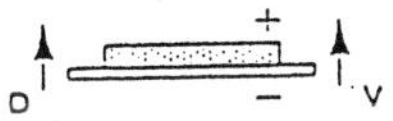	0.002	Expansion/ Contraction	P/E	10
Bimorph (bender)	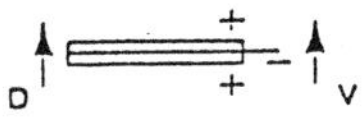	0.002	Expansion/ Contraction	P/E	20
Rainbow Monomorph (bender)	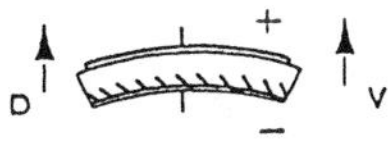	0.020	Expansion/ Contraction	P/E	35-500

Notes: V = Voltage; D = Actuator Displacement; P = Piezoelectric; E = Electrostrictor
* = Max. generated stress; + = Displacement at ± 10 kV/cm based on thickness

MATERIALS

Although a number of different compositions have been successfully prepared as Rainbows and Ferrofilms, those most compatible to the specific processes used and most amenable to achieving the desired properties are in the PLZT solid solution family. Typical high displacement ferroelectric compositions are 1/53/47 (La/Zr/Ti) and 5.5/56/44 for low and high dielectric constant applications, respectively; whereas, the usual compositions for the electrooptic, electrostrictive-type applications are 9/65/35 or 8.6/65/35.[9] These specific compositions are pointed out in the PLZT phase diagram given in Figure 1. As may be noted, the ferroelectric materials are morphotropic phase boundary compositions and the non-memory, electrostrictive materials are compositionally located along the ferroelectric-to-paraelectric phase boundary.

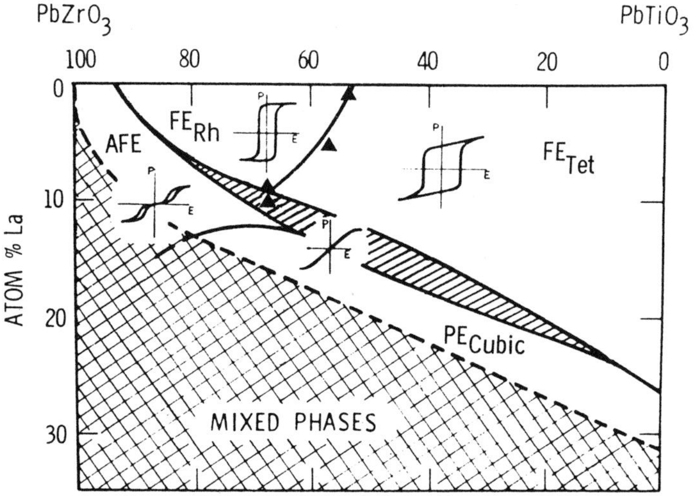

Figure 1. Room Temperature Phase Diagram of the PLZT System Compositions are indicated by triangular markers

PROCESSES

In this section, the specific processes which have been reported for both the Rainbows and the Ferrofilms are described. Since Rainbows are produced via a bulk-type process and the Ferrofilms are fabricated with a thick/thin film technique, they are distinctly different, by nature, and thus, are discussed in separate sections.

Rainbow Process

The Rainbow technology fundamentally consists of a new processing method that is applied to standard, high lead-containing ferroelectric, piezoelectric and electrostrictive ceramic wafers which are transformed by the process into a monolithic, composite structure consisting of a stressed dielectric and a chemically reduced, electrically conductive layer which acts as the stressing element as one of the electrodes for the final device. Since all of the materials are ferroelectric or electrically-enforced ferroelectric materials, they are multifunctional and smart, by nature, and are thus capable of performing both actuator and sensor functions, simultaneously.

The high temperature chemical reduction process involves the local reduction of one surface of a planar ceramic plate, thereby achieving an anisotropic, stress-biased, dome or saddle-shaped wafer with significant internal tensile and compressive stresses which act to increase the overall strength of the material and provide its unusually high displacement characteristics. According to previously reported work, the chemical reduction process proceeds via simple reactions consisting of the oxidation of the solid carbon (graphite) block to carbon monoxide and further oxidation of the carbon monoxide gas to carbon dioxide with the associated loss of oxygen from the PLZT oxide in contact or in near contact with the graphite block.

Rainbow ceramics are produced from conventionally sintered or hot pressed ceramic wafers by means of a few simple steps requiring approximately two hours of additional time as shown in the process flowsheet of Figure 2. A Rainbow is produced from an as-received wafer by placing it on a flat graphite block, placing a protective zirconia plate of the same size on top of the wafer and introducing the assembly into a furnace held at temperature in a normal air atmosphere. The part is treated at a temperature of 975°C for one hour, removed from the furnace while hot and cooled naturally to room temperature in about 45 minutes. A reduced layer approximately 150 um thick is produced in the wafer under these treatment conditions, When cool, the dome shaped wafer is lifted from the graphite block, brushed lightly on the reduced (concave) side to remove any metallic lead particles and to expose the reduced layer, and then electroded for test and evaluation. A variety of electrodes can be used such as epoxy silver, fired-on silver and vacuum deposited metals. After depositing appropriate electrodes, the Rainbow is completed and ready for operation.

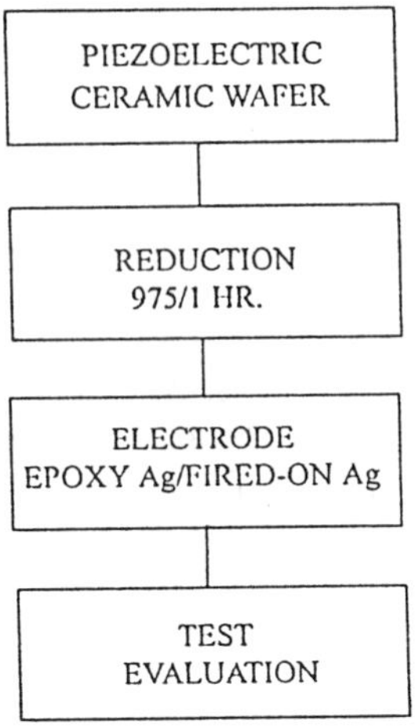

Figure 2. Process Flowsheet for PLZT Rainbow Ceramics

In regard to operation, a Rainbow is similar to a device known in the industry as a unimorph bender. A unimorph is composed of a single piezoelectric element externally bonded to a flexible metal foil which is stimulated into action by the piezoelectric element when activated with a ac or dc voltage and results in an axial buckling or displacement as it opposes the movement of the piezoelectric element. However, unlike the unimorph, the Rainbow is a monolithic structure with internal compressive stress bias on the piezoelectric element; thus producing the dome structure, rendering it more rugged and able to sustain heavier loads that normal. The integral electrode (usually the bottom electrode) consists of metallic lead intimately dispersed throughout the semiconductive oxide layer. The change in shape of the wafer after reduction is believed to be due to the reduction in volume of the bottom reduced layer (largely metallic lead) compared to the unreduced material, as well as the differential thermal contraction between the reduced and unreduced layers on cooling to room temperature.

Like other piezoelectric devices, Rainbows may be operated with a dc, pulse dc, or ac voltage; however, when driven with ac, the largest displacements are usually achieved at 100 Hz or less. In operation, the dome height of the Rainbow varies as a function of the magnitude and polarity of the voltage. When a given polarity of voltage is applied, the dome decreases in height depending on the magnitude of the voltage; and alternatively, when the polarity is reversed, the dome increases. The large axial motion of the dome is largely due to contributions from a lateral contraction produced in the material via the d_{31} coefficient and a stress-directed domain switching process.

It should be noted that although Rainbows are processed in bulk wafer form, after heat treatment they may be diced or scribed into smaller elements for a pick-and-place operation onto a hybrid substrate. This technique is possible since each individually diced element possesses a smaller but similar dome structure with a radius of curvature identical to the larger wafer. Even though the displacements of the smaller individual elements are proportionately less than the parent wafer, they nevertheless, are large enough (5 - 50 microns) to be useful in some devices. Some typical examples of sizes and shapes of Rainbows are shown in Figure 3.

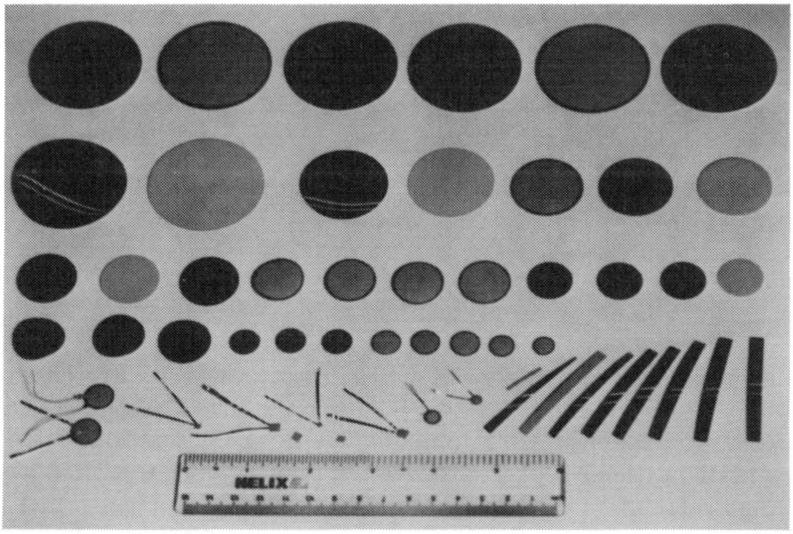

Figure 3. Typical Examples of Sizes and Shapes of Rainbows

Ferrofilm Process

Thick and thin films of ferroelectrics, piezoelectrics and electrostrictive materials are presently being fabricated from a water-soluble, acetate-precursor, liquid chemical system (Metal Organic Decomposition type) using an automated dipping process. An operational flowsheet for this process is given in Figure 4; however, the details of the process and the apparatus have previously been reported as part of a overall effort involving the intelligent processing of ferroelectric films.[10]

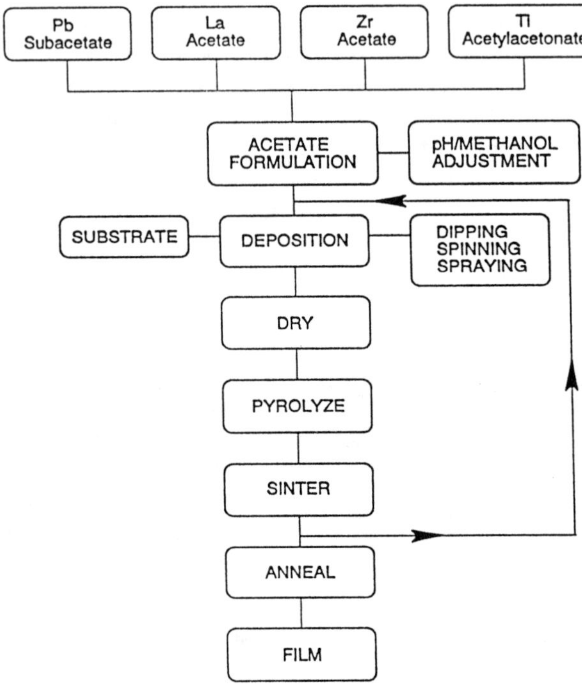

Figure 4. Process Flowsheet for PLZT Acetate MOD Process

Briefly, the process consists of formulating the stock solution from as-received precursors of Pb (lead subacetate), La (lanthanum acetate), Zr (zirconium acetate) and Ti (titanium acetylacetonate) by mixing them together in the requisite amounts along with the appropriate amount of methyl alcohol for viscosity control. This simple, five minute operation yields a clear, light yellow solution which is water soluble and stable for long periods of time. The solution is then deposited on the selected substrate via automatically controlled dipping and withdrawal operations. Drying occurs in a matter of a few seconds, and the coated substrate is subsequently sintered very quickly by introducing it directly into a furnace pre-heated to the sintering temperature. Multiple dipping, drying, sintering and cooling cycles are required in order to build up the necessary film thickness for the specific device. Depending on the dilution ratio of the solution, individually dipped layer thicknesses may vary from approximately 0.05 to 0.3 um, yielding films as thick as 12 um for a 40 layer device. For a cycle time of three minutes, the total time required to dip a 40-layer device is about two hours.

The final sintered film deposited on a suitable substrate such as a 0.125 mm thick Ag foil is usually transparent and crack-free with a smooth, shiny surface. After applying suitable electrodes such as air dried Ag, epoxy Ag or vacuum deposited metals, the film is ready for operation; however, it should be remembered that poling may be required if it is a ferroelectric thick film. Some examples of electroded and unelectroded thick films on Ag substrates are shown in Figure 5.

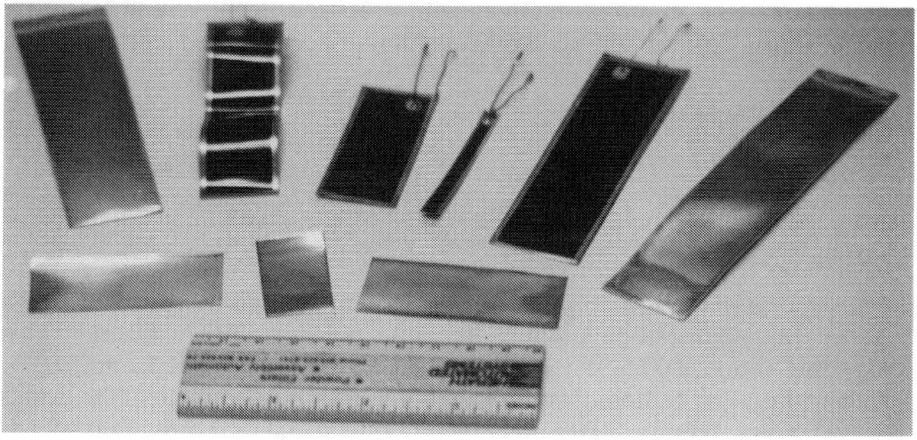

Figure 5. Typical Examples of Unelectroded and Electroded Ferrofilms

Although Ferrofilms lend themselves well to a number of different fabrication processes at the totally integrated level, for discrete components at the hybrid microelectronics level the dipping technique has been found to be reliable and predictable and can easily be implemented in a totally hands-off environment. In addition, industrial equipment is readily available to dip very large as well as small, discrete parts at minimal cost. The acetate process, being water insensitive, is especially suitable for the dipping process since the open solution is usually exposed to the atmosphere for long periods of time and during this period must withstand chemical interactions as well as minimal evaporation of solution.

PROPERTIES

In the last two years since the Rainbow ceramics were first developed at

Clemson University, there has been a considerable on-going effort to (1) understand the details of the reduction process in the PLZT materials, (2) measure their properties and characterize their unusually high displacement and load bearing capabilities as benders, (3) model their electromechanical behavior and frequency dependent properties and (4) construct working models illustrating various proof-of-principle applications. Some of this data for the Rainbow ceramics will be presented in this section, along with the limited amount of data available, to date, for the Ferrofilms. In most instances, a standard Rainbow size of 31.75 mm diameter and 0.5 mm thick was used to obtained the data. Additional technical data on the Rainbow ceramics has been reported in a document prepared by Sherrit, et.al., of the Royal Military College of Canada.[11]

Rainbows Characteristics

Dielectric Properties - The temperature dependent dielectric behavior for two PLZT compositions; i.e., 1/53/47 and 8.6/65/35, are shown in Figures 6 and 7, respectively. It can be seen from Figure 6 that a gradual rise occurs in the relative dielectric constant (1 kHz) of 1/53/47 from a room temperature value of approximately 1100 to about 2700 at 200°C. No peak is observed in this range for this composition because its Curie point is 330°C. On the other hand, composition 8.6/65/35 in Figure 7 shows a change in K from 3200 to 5700 over this same temperature range with a peak occurring at 105°C, which is its usual Curie point as determined from small signal measurements. Since this composition is an electrostrictive, relaxor-type material, this Curie point does not coincide with its loss in polarization which occurs at about 20°C; thus, making it one of the most sensitive, high displacement, electrostrictive Rainbow materials. It may be noted that the dielectric constants and dissipation factors for both compositions are comparable to previously reported values, and this indicates that the Rainbow reduction process does not change the dielectric properties of the unreduced part of the structure.

Hysteresis Loops - Typical examples of dc hysteresis loops for compositions 1/53/47 and 9/65/35 are given in Figure 8. The loop in Figure 8(A) was taken on the ferroelectric Rainbow element (1/53/47) in its virgin condition before any other measurements were made. It should be noted that on the initial application of positive voltage to +450V there was approximately 60% of the total remanent polarization switched rather than the usual 50% one ordinarily observes in a virgin, randomly oriented ceramic.. This behavior is highly unusual and indicates that the Rainbow ceramic was partially poled before testing. Additional audio and piezoelectric tests of other virgin parts also indicated that the elements were partially poled to varying degrees; i.e., some very little and others as high as 75%.

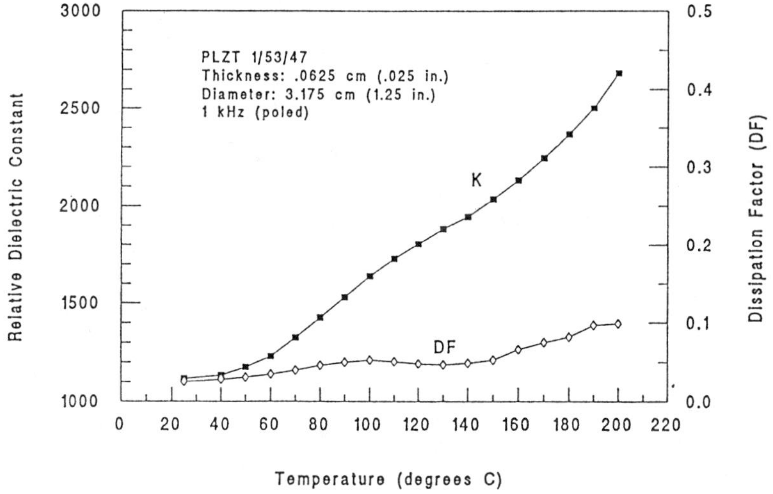

Figure 6. Temperature Dependent Dielectric Properties of PLZT Rainbow 1/53/47

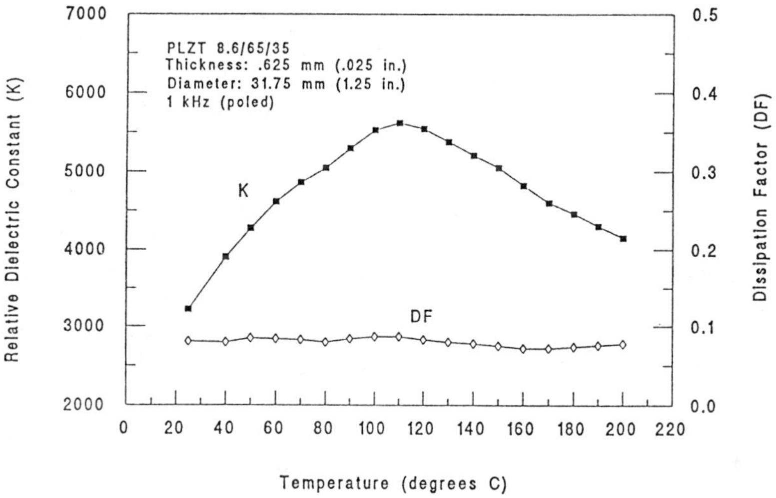

Figure7. Temperature Dependent Dielectric Properties of PLZT Rainbow 8.6/65/35

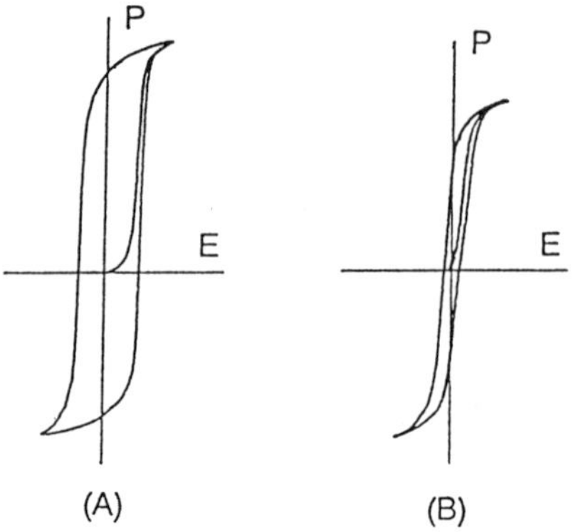

Figure 8. Virgin Hysteresis Loops for Rainbow PLZT Compositions (A) 1/53/47 and (B) 9/65/35

One explanation for this condition occurring in the electrically virgin state is that the mechanical compressive and tensile stresses produced in the Rainbow wafer during processing are acting together to switch some of the domains in this soft ferroelectric/ferroelastic material. Since uniform stress is a symmetrical quantity, it is recognized that it alone is insufficient to produce a net polarization in a given direction even though it may be of sufficient magnitude to switch domains; however, a stress gradient such as produced by the Rainbow bending process is a vector quantity and can, indeed, produce the observed effect. This non-uniform stress is believed to be responsible for the partial poling of the Rainbow wafers. Measured properties on the above wafer were: P_R = 44.8 uC/cm^2, E_C = 7.5 kV/cm, dielectric constant = 1210 and dissipation factor = 0.047.

The virgin loop of Figure 8(B) is a typical one for the electrostrictive (9/65/35) type of Rainbow materials and is very similar to that obtained on bulk electrooptic material. Measured properties on this wafer were: $P_{10kV/cm}$=28.3 uC/cm^2, dielectric constant = 3142 and dissipation factor = 0.085. As a matter of course, no unsymmetrical hysteresis loops were observed in the electrostrictive materials, and none was expected, since there are no stable domains in these materials at zero electric field. Conceivably, a high enough stress could precipitate stable domains in a very near-ferroelectric material, however, this was not experimentally confirmed

Displacement Loops - Displacement vs. electric field (butterfly) loops for the Rainbow wafers described above are shown in Figure 9. As before, Figure

9(A) illustrates the Rainbow axial motion as the sample is electrically switched from zero to +450V, to -450V and back to zero, however, in this case this loop was not taken on the virgin wafer. It may be noted that this loop is remarkably similar to that observed when measuring the direct extensional (longitudinal, lateral) displacements via the piezoelectric d_{33} or d_{31} coefficients. The value of displacement in the + voltage direction was measured at 190.5 um, and the total amount of displacement (+/-) was 432 um.

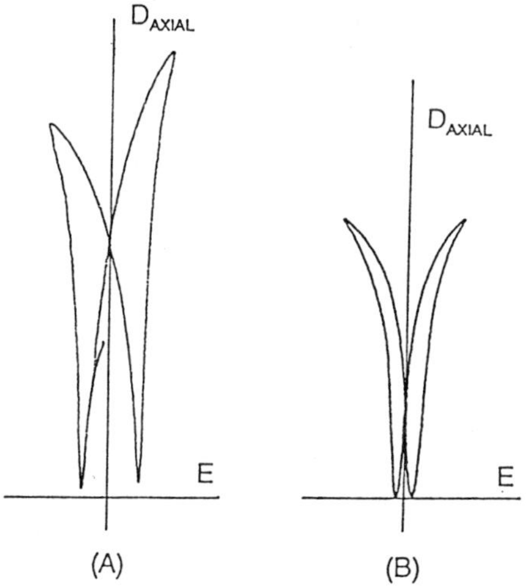

Figure 9. Axial Displacement Loops for Rainbow PLZT Compositions (A) 1/53/47 and (B) 9/65/35

Figure 9(B) shows the displacement loop of the electrostrictive Rainbow material (9/65/35) mentioned above. Since 9/65/35 is a relaxor material there should be little or no memory, and the same value and sign of displacement should be obtained whether a + or a - voltage is applied. One can see by switching this sample through a full voltage loop that a small amount of remanent displacement (strain) is present which is probably due to the close proximity of this composition to a FE phase. A further indication of this incipient FE phase is the higher than normal value of P_{10} (P_{10} = 28.4 vs. 18.0 uC/cm^2) as given above. Measured value of total displacement for this wafer was 178 um.

Displacement Characteristics - The displacement characteristics as a function of applied voltage are given in Figure 10 for some selected compositions. One of the most striking features of this figure is the very high displacements achieved by these Rainbow ceramics at moderate electric fields; e.g., 400 volts is equivalent to an electric field of 10 kV/cm. Composition 8.6/65/35 is noted to possess the highest displacement of 210 um at a maximum voltage of 600 volts, however, its displacement is characteristically non-linear because of its electrostrictive nature. Compositions 1/53/47 and 5.5/56/44 are ferroelectric materials, and thus, are more linear in behavior. As a general rule, the displacements of the ferroelectric materials are lower than those of the electrostrictive compositions, particularly when operated at higher voltages and one polarity; however under bipolar operation, the displacement values of theferroelectric materials will commonly be double the values shown in Figure 10.

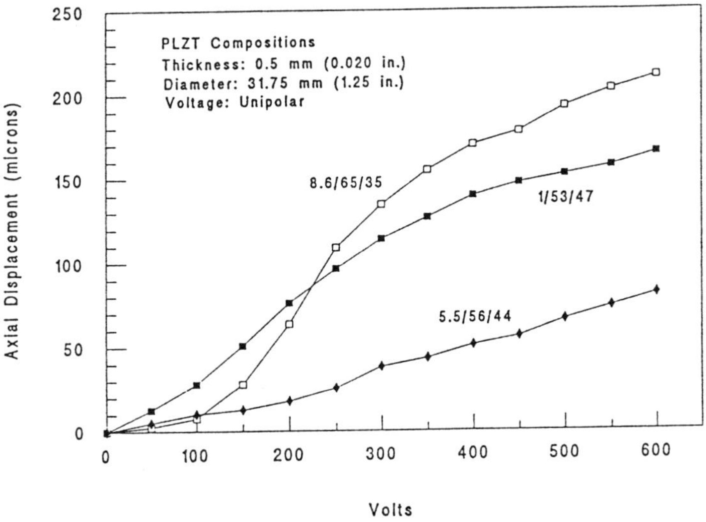

Figure 10. Axial Displacement Characteristics of Rainbow PLZT Compositions as a Function of Voltage

A measure of the difference in displacement between the planar (lateral extensional mode) direction and the orthogonal axial (Rainbow bender mode) direction is given in Figure 11 for PLZT 8.6/65/35, which also shows the temperature dependence of these two modes. This figure clearly demonstrates the very large displacement amplification of the bending phenomenon when one considers the data showing a change in displacement from approximately 0.07% to 22% at 25°C; i.e., an amplification of 315%. Although not shown on the figure, both displacements are negative (i.e., a contraction) when voltage is applied.

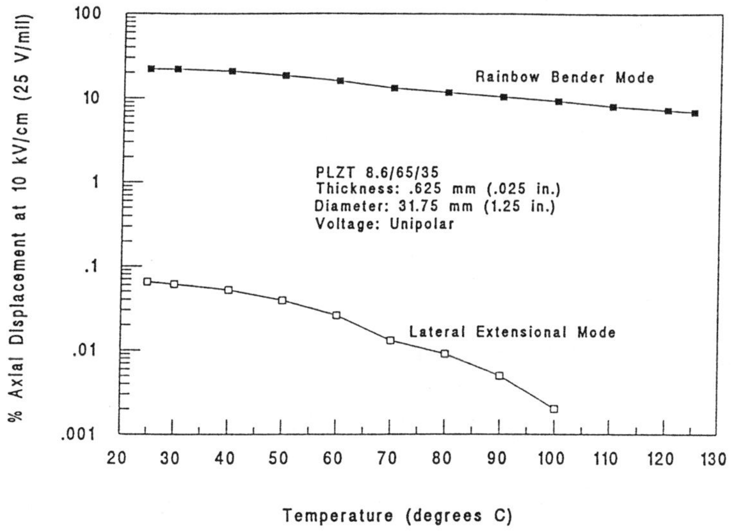

Figure 11. Temperature Dependence of Electrostrictive Lateral and Rainbow Bending Displacements in PLZT 8.6/65/35 Ceramics

Figure 12 illustrates the profile of the vertical bending displacement across the diameter of a Rainbow wafer during activation at 10 kV/cm. As might be anticipated, the highest displacement is in the center of the wafer, dropping off to zero at the circumference. This zero displacement at the edge of the wafer is beneficial because a Rainbow can be conveniently placed on a planar surface and operated as a linear actuator device or its circumference can be sealed off, and it can be operated as a cavity-mode pump.

The variation of a Rainbow's axial displacement as a function of wafer diameter is given in Figure 13. For a 0.5 mm thick wafer of composition 1/53/47, the values of displacement can be seen to vary from 170 um for a diameter of 31.75 mm (1.25 inch) to approximately 3 um at 6.5 mm diameter. Thus, a discrete 15 mm diameter Rainbow component on a hybrid substrate could be expected to have a displacement of about 40 um when operated with a single polarity and about 80 um when operated bipolar.

Wafer thickness has been found to have a significant effect upon axial displacement primarily because of the change in motional mode; i.e., from dome to saddle-type, as the wafer thickness is reduced to approximately one one-hundredth of the diameter. For example, a 31.75 mm (1.25 inch) diameter wafer usually develops a saddle-mode configuration when its

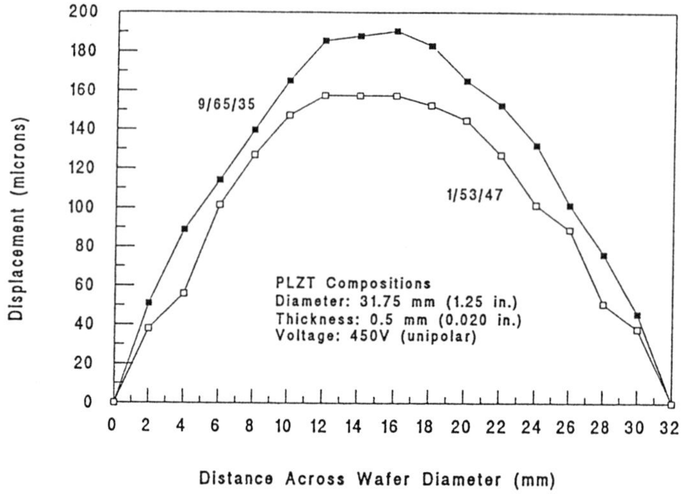

Figure 12. Profile of Displacements Measured Across Rainbow Wafer Diameter for PLZT 1/53/47 and 9/65/35

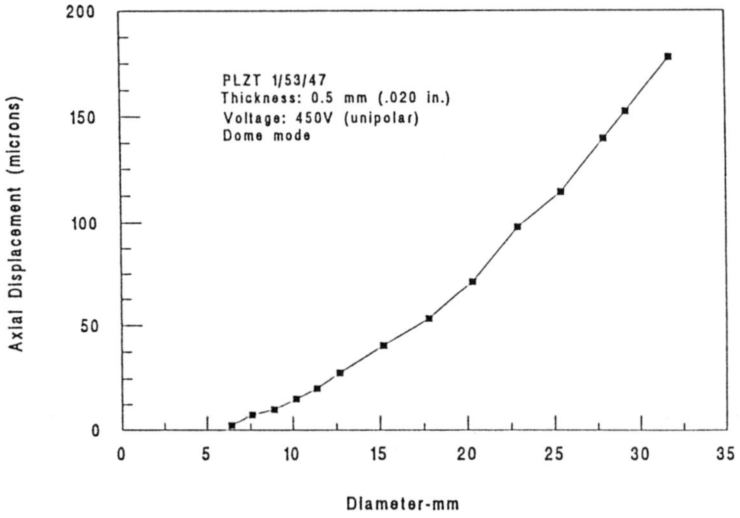

Figure 13. Variation of Axial Displacement as a Function of Rainbow Wafer Diameter for PLZT 1/53/47

thickness is less than 0.32 mm (0.013 inch). Saddle-mode operation provides maximum displacement with minimum load bearing capability; and therefore, should only be considered for special applications. Figure 14 illustrates the unusually large range of displacements obtained for Rainbows as a function of thickness. Please note that the displacement axis is a log scale.

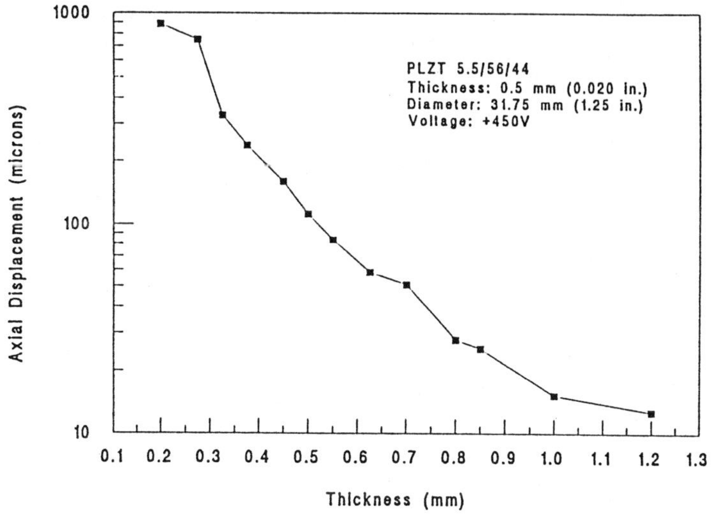

Figure 14. Effect of Thickness on Axial Displacement for PLZT 5.5/56/44

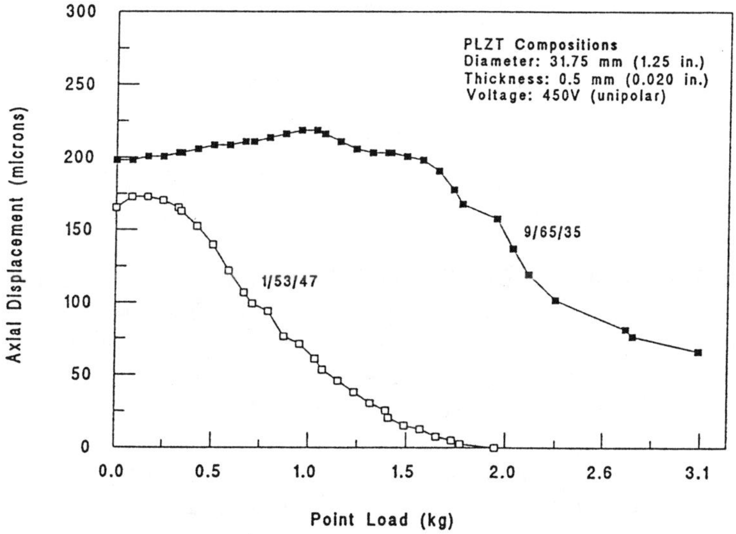

Figure 15. Point Load-Bearing Properties of Rainbow PLZT Compositions

The effect of an unconstrained point load on the displacement of an activated Rainbow is given in Figure 15 for compositions 1/53/47 and 9/65/35. PLZT 1/53/47 can be seen to be relatively ineffective when loaded with a dead weight of 1.5 kg (3.3 lbs), whereas, composition 9/65/35 is still effective at a load of over 3 kg. This result is not too surprising since the elastic modulus of 9/65/35 (10.9 x10^4 MPa) is noticeably higher than that of 1/53/47 (5.7 x 10^4 MPa). Another point to note from the figure is the increase in displacement with the introduction of a finite amount of load on the device. This effect was previously reported by Furman, et. al.[12] The tradeoff between thickness and maximum sustainable point loading is given in Figure 16 for 1/53/47.

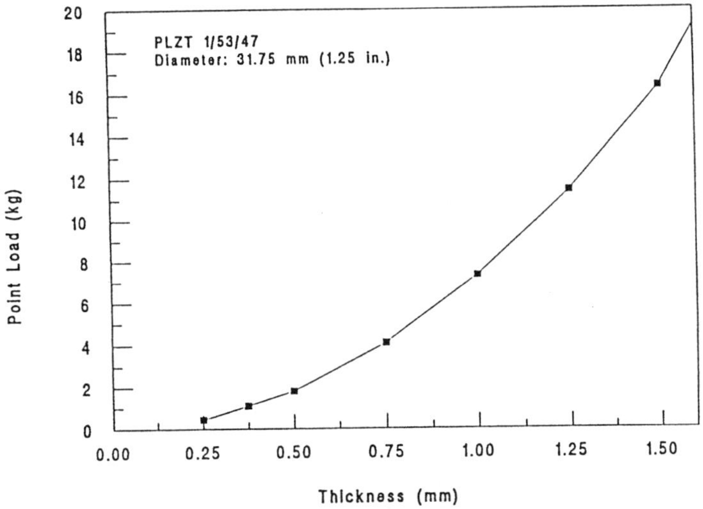

Figure 16. Maximum Sustainable Point Load for Rainbow PLZT 1/53/47 as a Function of Thickness

Another concern of actuator designers is the amount of force that can be generated by an actuator when voltage is applied. This is shown in Figure 17 for a 1/53/47 Rainbow of standard size. As can be seen, the force generated is a linear function of voltage until the onset of saturation for this particular configuration. A maximum force of 1.3 kgf was achieved at 450 volts.

Finally, displacement and sustainable overpressure data for PLZT 5.5/56/44 as a function of wafer thickness are presented in Figure 18. The left ordinate scale represents the maximum allowable pressure differential across the wafer thickness before the wafer contacts the planar surface and stops flexing

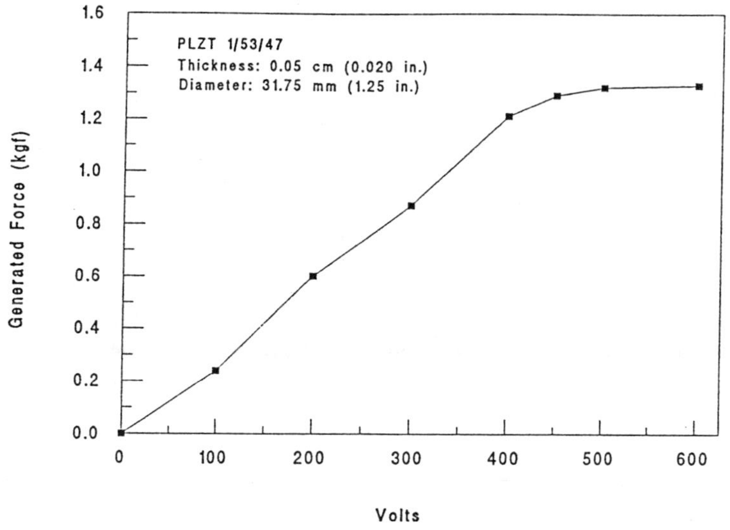

Figure 17. Effect of Applied Voltage on the Force Generated by a PLZT 1/53/47 Rainbow Wafer

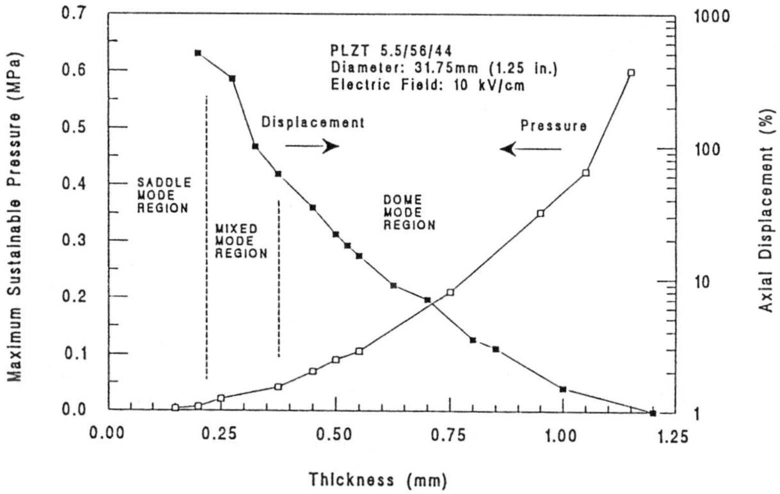

Figure 18. Sustainable Pressure and Displacement Characteristics for Rainbow PLZT 5.5/56/44 as a Function of Thickness

usually <0.5 mm thickness) or the wafer mechanically fractures (>0.5 mm thickness). The right ordinate log scale is the percent displacement (based on wafer thickness) of the device when operated at 450 volts dc while only under the loading of the dial gage (80 grams). Also indicated on the figure are the wafer thickness regions where the saddle and dome modes of operation are dominant for a wafer 31.75 mm (1.25 in.) in diameter. The data shown in the figure indicate that Rainbow displacements span an unusually large range from near zero to at least 500% with actual displacements of up to 1 mm (0.040 in.) for a 0.2 mm (0.008 in.) thick wafer. Of course, such large displacements are not possible when operating under significant pressure differentials or under moderate point loading situations near its capacity.

Resonance Characteristics - The resonant, frequency dependent properties of a standard Rainbow wafer is given in Figure 19 which displays both impedance and phase angle as a function of frequency. Some outstanding features of this figure are the large radial resonance anomaly at 70 kHz and the several bending resonances between 1 to 20 kHz. Other resonances not shown in this figure are (1) overtone resonances of the fundamental radial resonance in the range from 100 kHz to 1 MHz, (2) thickness resonances in the low megahertz range around 4 and 8 MHz for the fundamental and first overtone and (3) very low frequency structurally-dependent resonances in the range of 25 to 500 Hz which can be noted when a Rainbow is operated as part of a working device.

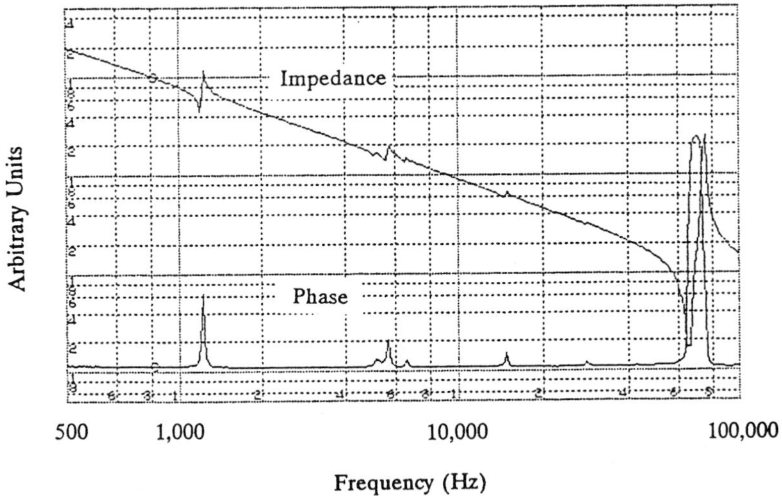

Figure 19. Resonance Characteristics of a Rainbow PLZT 1/53/47 Wafer

Table of Properties - A summary of selected properties which have been determined, to date, for the Rainbow ceramics are given in Table 2. In addition to these, a number of advantages and features of the Rainbows are: (1) simplicity, (2) solid-state type, (3) monolithic, (4) pre-stressed for greater strength and durability, (5) can sustain/generate moderate loads, (6) mechanical overload protection, (7) medium fast response, (8) very high axial displacement, (9) surface mountable, (10) above-the-plane displacement, (11) no bonding layers, (12) temperature compensation possible, (13) can be stacked to multiply strain, (14) easy to fabricate and (15) cost effective.

Table 2. Some Selected Properties of Rainbow Ceramics

Thermal Expansion -	10.0×10^{-6}/°C (reduced layer) 5.4×10^{-6}/°C (oxide layer)
Modulus of Rupture -	1.58×10^{2} MPa (22,500 psi)
Modulus of Elasticity -	5.7×10^{4} MPa (8.1×10^{6}psi)(bending)(1/53/47) 10.9×10^{4} MPa (15.5×10^{6} psi)(bending)(8.6/65/35)
	6.9×10^{4} MPa (reduced layer)(acoustic)(5.5/56/44) 7.8×10^{4} MPa (oxide layer) (")(")
Acoustic Velocity -	4015 m/sec (reduced layer) 4248 m/sec (oxide layer)
Poisson's Ratio -	0.38 (reduced layer) 0.38 (oxide layer)
Resistivity -	3.8×10^{-4} ohm-cm
Effective $d_{33\,(saddle)}$ -	2.5×10^{-6} m/V (bending mode)
$d_{33\,(dome)}$ -	2.8×10^{-7} m/V (bending mode)
Hysteresis -	5 - 35%
Displacement (31.8 mm) -	178 microns (0.007 inch)(0 - 450V)(dome mode) 381 microns (0.015 inch)(+/- 450V)(dome mode) 1143 microns (0.045 inch)(0 - 450V)(saddle mode)
(102 mm dia.) -	3175 microns (0.125 inch)(0 - 450V)(saddle mode) 1016 microns (0.040 inch)(0 - 450V)(dome mode)
Capacitance (1.25" dia.) -	15 nF (PLZT composition 1/53/47) K = 1200 30 nF (PLZT composition 5.5/56/44) K = 2400 60 nF (PLZT composition 9/65/35) K = 3800

Ferrofilms

The properties of Ferrofilms were determined from PLZT composition 2/53/47 prepared via the acetate process and automatically dip coated for 40 layers onto Ag substrates ranging in thickness from 0.025 to 0.25 mm, yielding an overall film thickness of approximately 12 um. Standard, 1-mm diameter, vacuum evaporated Cu electrodes were deposited on the surface of the films for testing.

Dielectric Properties - The small-signal dielectric measurements of the PLZT thick films revealed that they possess properties very similar to bulk material of the same composition. For the particular composition evaluated, relative dielectric constants (1 kHz) ranged from 1400 to 1800 and dissipation factors from 3.5% to 4.6%.

Hysteresis Loops - Hysteresis loops (1 KHz) were obtained with an ac looper constructed in-house because thick films require higher voltages (>100 V) than normally available from standard thin film testers. An example of a typical loop is shown in Figure 20. As can be seen, the loop (polarization vs. electric field) is very square with sharp, saturated loop tips at maximum field, indicating a high degree of domain switching and good, insulating characteristics well into saturation at 100 kV/cm. The loop of Figure 20 displays typical properties for these films; i.e., that of: P_R = 42 uC/cm^2 and E_c = 15.7 kV/cm (40 V/mil). It should be recognized that this low value for E_c is more typical for bulk material than for thin films which characteristically possess E_cs of 75 kV/cm or higher.

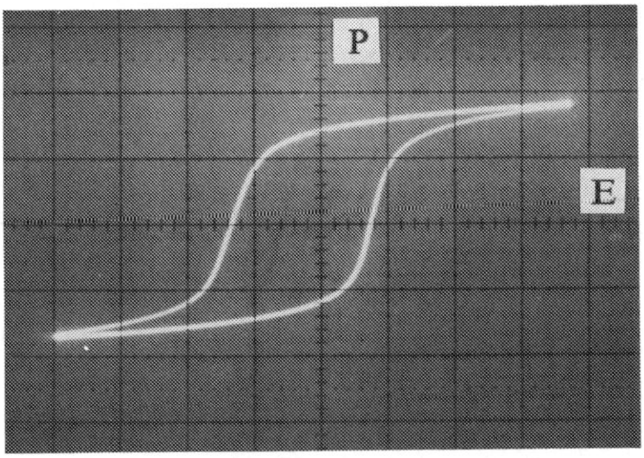

Figure 20. Typical Hysteresis Loop of Ferrofilm PLZT 1/53/47 on Ag

APPLICATIONS

A number of examples of applications are given in this section in order to demonstrate the versatility of the Rainbow and Ferrofilm technologies. These working models are essentially discrete, proof-of-principle devices which require further engineering design, miniaturization and modification in order for them to be suitable for hybrid microelectronics or integrated structures. In any case, it is believed that the Rainbow technology best serves the application range from macroelectronics to millielectronics, whereas the thick/thin Ferrofilms are best suited for the range from millielectronics to microelectronics.

The Rainbow devices shown in Figures 21 and 22 are typical examples of a number of applications envisioned for this technology. As can be seen, they range from actuators to sensors, and speakers to pumps. A more extensive list of applications include (1) linear actuators, (2) cavity/piston pumps, (3) loud speakers, (4) reciprocating motors, (5) relays/switches/thermostats, (6)

Figure 21. Examples of Working Model Devices Using Rainbow Ceramics as Actuators

sensors, (7) hydrophones/hydroprojectors, (8) variable-focus mirrors/lenses, (9) optical deflectors/scanners, (10) vibrating delivery systems, (11) liquid delivery systems, (12) antivibration/noise cancelling devices, (13) displays, (14) sonic and ultrasonic devices and (15) auto-leveling platforms.

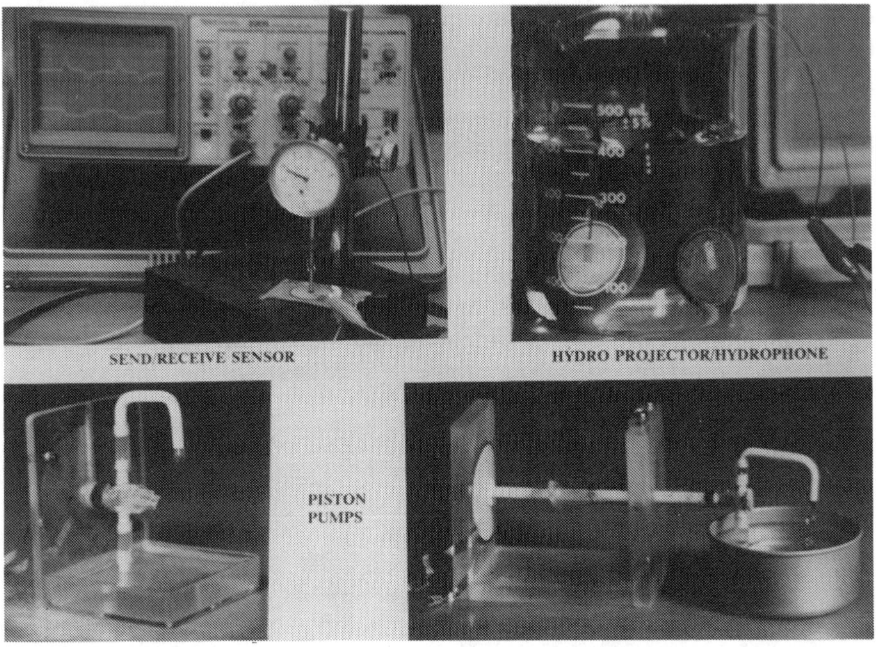

Figure 23. Additional Examples of Working Models Using Rainbow Ceramics as Actuators and Sensors

A single Ferrofilm device is shown in Figure 23. It simply consists of a 25 x 50 mm x 12 um thick film which was dip coated onto a 0.125 mm thick Ag substrate and electroded on the top major surface with vacuum-deposited Cu. After poling the film at 70 volts, the film/substrate is mounted to a resonating enclosure (in this case the enclosure is a cardboard box) and connected to the output of a radio. As with most ferroelectric/piezoelectric audio devices, the quality of the audio is only moderate, at best, when operated over the full audio range of the radio.

CONCLUSIONS

The prospects for utilizing Rainbows and Ferrofilms in discrete hybrid and totally integrated microelectronics are promising for future applications involving smart ceramics such as ferroelectrics, piezoelectrics, pyroelectrics, electrooptics and electrostrictive materials. Rainbows have opened up a new dimension in ultra-high displacement actuators while Ferrofilms have now bridged the gap from bulk materials to thin films. The key to adapting these materials to specific devices and applications is the manner in which answers are found to questions concerning their reproducibility, reliability, longevity and cost effectiveness. Further development and design work are obviously needed in order to answer these questions.

Figure 23. A Ferrofilm Speaker for a Stereo System

REFERENCES

1. L.H. Sheppard, "Advances in Processing of Ferroelectric Thin Films, Bull. Am. Ceram. Soc., 71, No. 1, 85-95 (1992)

2. P.C. Fazan, "Trends in the Development of ULSI DRAM Capacitors," Integrated Ferroelectrics, 4, 247-256 (1994)

3. R. Watton, "Ferroelectric Materials and IR Bolometer Arrays: From Hybrid Arrays Towards Integration," Integrated Ferroelectrics, 4, 175-186 (1994)

4. E.M. Lenoe, W.N. Radicic and M.S. Knapp, "Implications of Smart Materials in Advanced Prosthetics," Proceedings SPIE on Smart Structures and Materials, 2189, 84-104 (1994)

5. C.C. Hsueh, T. Tamagawa, C. Ye, A. Helgeson and D.L. Polla,"Sol-Gel Derived Ferroelectric Thin Films in Silicon Micromachining," Integrated Ferroelectrics, 3, 21-32 (1993)

6. G.H. Haertling, "Rainbow Ceramics - A New Type of Ultra-High Displacement Actuator," Bull. Am. Ceram. Soc., 73, No. 1, 93-96 (1994)

7. G.H. Haertling, "Chemically Reduced PLZT Ceramics for Ultra-High Displacement Actuators," Ferroelectrics, 154, 101-106 (1994)

8. K.K. Li, G.H. Haertling and W-Y Howng, "An Automatic Dip Coating Process for Dielectric Thin and Thick Films," Integrated Ferroelectrics, 3, 81-91(1993)

9. G.H. Haertling, "Dielectric and Electrooptic Properties of Acetate Derived PLZT X/65/35 Thin Films," Integrated Ferroelectrics, 3, 207-215 (1993)

10. G.H. Haertling, "Intelligent Processing of Ferroelectric Films," Bull. Am. Ceram. Soc., 73, No. 8, 68-73 (1994)

11. S. Sherrit, H.D. Wiederick and B.K. Mukherjee, "Evaluation of PLZT Based RAINBOW Ceramic Samples Developed by Dr. Gene Haertling at Clemson University," Report No. DREA CR/94/436, Royal Military College of Canada, Kingston, Ontario, Canada, June (1994)

12. E. Furman, G. Li and G.H. Haertling, "Electromechanical Properties of Rainbow Devices," Proceedings of the 9th ISAF Meeting, State College, June (1994)

SILICON-ELECTROCERAMICS INTEGRATION

Harry L. Tuller
Crystal Physics and Electroceramics Laboratory
Department of Materials Science and Engineering
Massachusetts Institute of Technology
Cambridge, MA 02139

ABSTRACT

Electroceramics provides unique functionality as sensors, transducers, and non-volatile storage media while silicon is nearly unique in its signal processing and micromechanical properties. We discuss how these two materials technologies, when integrated, provide unique devices and structures.

INTRODUCTION

The unique characteristics of ceramics such as their thermal, chemical, and mechanical stability have been of practical interest to mankind since at least the 6th century B.C. [1]. Optically transparent and colored glasses date from approximately the same period [2]. Technological uses of electroceramics came much later, for example, the construction of the spark plug by J.J. Lenoir of France in 1860 [3]. The use of ceramics as electrical insulators expanded into the first half of the twentieth century in the form of high voltage standoffs, vacuum tube insulators, and glass-metal seals. In all these uses, ceramics were admired primarily for their passive characteristics.

The modern era of electroceramics, characterized by the discovery and application of ceramics with exceptional or active electrical, optical or

To the extent authorized under the laws of the United States of America, all copyright interests in this publication are the property of The American Ceramic Society. Any duplication, reproduction, or republication of this publication or any part thereof, without the express written consent of The American Ceramic Society or fee paid to the Copyright Clearance Center, is prohibited.

magnetic properties, was ushered in during the middle of this century. Particularly notable examples include the discovery and development of $BaTiO_3$, in the 1940's and early 1950's, as a ferroelectric and piezoelectric material [4] and the spinel ferrites as magnetic materials [5]. Both of these classes of materials have since led to the development of key device technologies including multi-layer capacitors, PTC thermistors, electronic filters, electro-mechanical transducers, magnetic recording heads and media, permanent magnets, and electro-optic modulators.

Over the last several decades, there has been worldwide recognition that steps must be taken to conserve our natural resources and limit emissions of pollutants and toxic chemicals into the environment. This, among other things, requires that we learn to control and improve the efficiency of chemical processes including combustion control which accounts for a major source of both fuel consumption and air pollution. The proliferation of inexpensive microprocessors now makes on-line monitoring and control of complex processes possible, including those important for medical and consumer applications. The key to the feasibility of such closed-loop systems is the existence of suitable electrochemical, mechanical, and thermal sensors and actuators which serve as the input and output functions for the microprocessors (see Fig. 1).

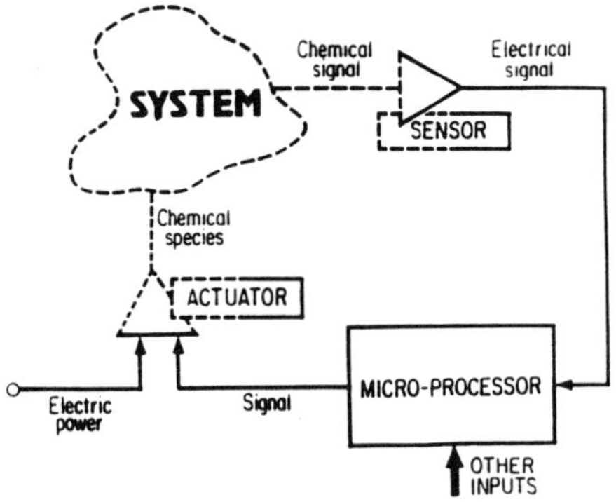

Figure 1. Schematic of a feedback loop with a chemical sensor and actuator as an example [6].

Polycrystalline ceramics are ideally suited for serving as the sensor material in many of these applications. First, they exhibit excellent stability over wide operating limits. They can, for example, withstand the high temperatures and aggressive chemical environments characteristic of automotive and industrial environments. Perhaps, more importantly, as a class of materials, they exhibit a great deal of versatility in properties which make them ideal transducer materials. These include, but are not limited to:

- Electronic conduction (atmosphere and temperature sensitive)
- Fast ion conduction
- Ferro-, Piezo, and Pyro-electricity
- Ferro-magnetism
- Electro-optic modulation
- Luminescence

This versatility leads to many opportunities for the development of key classes of sensors and actuators including those with chemical, thermal, mechanical, magnetic, electrical, and optical functions [6]. When one combines this with flexibility in processing (bulk, thick, and thin films) and low cost, the attraction of using electroceramic material in these and related applications becomes apparent [6].

There are, however, longstanding trends in the field of electronics which are beginning to impact related technologies. These include the incessant drive towards miniaturization, enhanced functionality and batch processing of devices. These are all characteristics of silicon-based integrated circuit technology within which rapid improvements in device capabilities and reduced cost per bit are manifest.

Let us begin by addressing the question of enhanced functionality. The utility of sensors and actuators in a feed-back scheme of the type depicted in Fig. 1 is considerably enhanced if the sensing, signal processing and actuation functions can be integrated onto a single chip. Since signal processing is now largely performed at low cost in silicon-based devices, this suggests the need for integrating the materials used for the sensing and actuating functions onto the silicon chip. Enhanced functionality for a sensor-based device may also reflect the need for enhanced selectivity and/or reduced response time for detection of a specified gaseous species or the ability to monitor a number of parameters simultaneously, e.g., temperature, pressure, and concentration of a chemical species. Such "smart sensors" obviously require a silicon chip in the system.
While conventional bulk ceramic sensors provide the versatility and low cost described above, they typically suffer from large size, high power consumption, reproducibility, and poor selectivity in distinguishing between different chemical species. Without integration, subsequent signal processing by costly stand-alone equipment becomes necessary.

Until the mid to late 1980's, there was little incentive for silicon technologists to incorporate complex oxides and other refractory compounds into their silicon "fabs". For example, fears associated with contamination of silicon by transition metals such as Cu or Fe, biased processing engineers away from considering new materials. Several events in the latter half of the past decade have induced a change in attitude. The most significant being the discovery of high temperature superconductors. Many groups became interested in applying this technology for memory storage (Josephson junctions), sensing (magnetic fields, temperature), microwave circuits, interconnects, etc. Many were able to quickly demonstrate that it was possible to deposit high quality complex cuprate films on silicon wafers by selecting one of a number of low temperature processing schemes for the oxides, e.g., laser ablation, sputtering, sol gel chemistry, and by laying down appropriate intermediate buffer layers. There has been a significant growth of interest in ferroelectric-based nonvolatile memories relying on the use of perovskite or perovskite-related materials such as lead zirconate-titanate (PZT). Here again, progress has been rapid in demonstrating the feasibility of integrating complex functional oxides onto silicon chips.

To conclude, the field of microelectromechanical (MEMS) devices is a rapidly growing silicon-based technology which serves as an additional incentive for the integration of ceramic materials and silicon devices. This technology uses chemical etching to produce the planar architecture characteristic of integrated circuits, thereby creating more three-dimensional-like structures such as microcantilever beams, bridges, and membranes. A number of new high volume applications have followed in the development of compact pressure sensors and accelerometers. Also, micro-pumps and valve systems, scanning force microscope heads, micro-mirror video arrays, etc., have either arrived on the market in smaller quantities or are still under development.

The significance of MEMS technology to our discussion is the following. There is the ability to substantially shrink present sensor and actuator devices and to fabricate them by batch processing, thereby creating smaller, rugged and potentially more sensitive devices at a lower cost. Although not obvious, there is the ability to integrate small three dimensional structures with the microcircuitry necessary for signal processing. Lastly, by combining electroceramic films with these MEMS structures, additional interesting synergisms become possible (see Fig. 2).

In the following, we provide examples of how silicon-electroceramic integration provides opportunities for new and/or improved devices, and discuss some of the difficulties that must be overcome. We also discuss some of our recent work on photoelectrochemical machining of silicon which provides additional flexibility in achieving stress-free, high aspect ratio MEMS structures.

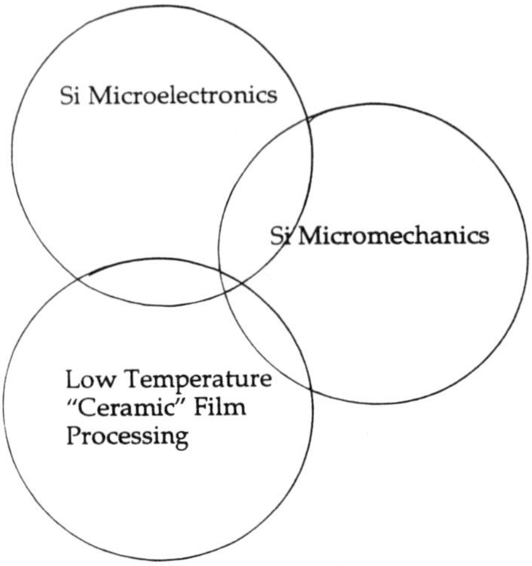

Figure 2. The integration of silicon microelectronics, micromechanics, and electroceramic films opens up new opportunities for device technology.

CHEMICAL SENSORS

As illustrated in Fig. 1, appropriate chemical sensors and actuators are essential components of feedback systems designed to monitor and control the chemical state of a system. Generally, desirable attributes of chemical sensors include (1) high sensitivity (2) selectivity (3) long-term stability and (4) low cost. In many cases, (5), rapid response times are also required, e.g., as in the closed-loop-control of the air-to-fuel ratio (A/F) in automobiles [6].

A number of sensing principles can be applied using electroceramics which rely either on their dielectric, semiconducting, or electrolytic properties. The first of these to be discussed is one which utilizes a device structure, the metal-oxide semiconductor field-effect-transistor (MOSFET), which has long utilized silicon-electroceramic integration [7].

A schematic of the cross section of an n-channel MOSFET device is reproduced in Fig. 3. Normally, no current flows between the n^+ source (S) and drain (D) electrodes due to the reversed biased p-n junction between the drain and base (B). An n-type conductive channel can, however, be induced by the application of a positive bias on the gate (G) electrode thereby inducing a carrier inversion (p to n) just below the SiO_2 layer. The gate voltage V_G necessary to "turn-on" the device is called the threshold voltage V_T. The output current-voltage characteristics take the form shown in Fig. 4, in which one sees that the drain current, I_D, remains negligibly small for $V_G<V_T$, while increasing with increasing $V_G>V_T$, due to an increasing channel width. The saturation current follows a square-law dependence [8] given approximately by

$$I_{Dsat} = \frac{Z\bar{\mu}_n C_0}{2L}(V_G - V_T)^2 \qquad (1)$$

where Z and L are the channel depth and length, $\bar{\mu}_n$ the effective electron mobility in the channel and C_o, the oxide capacitance.

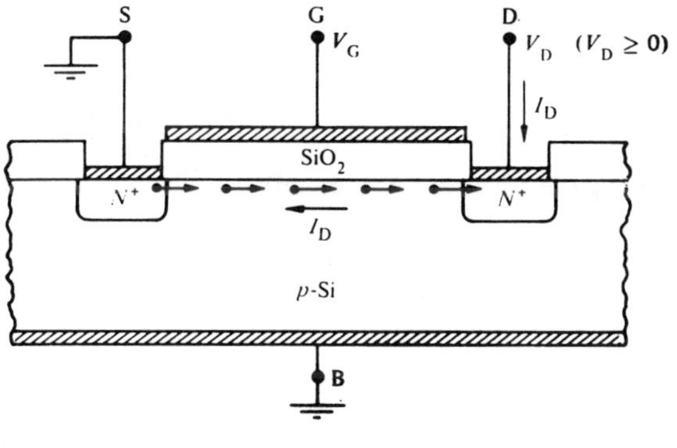

Figure 3. Schematic cross section of an n-channel MOSFET. From Pierret [8].

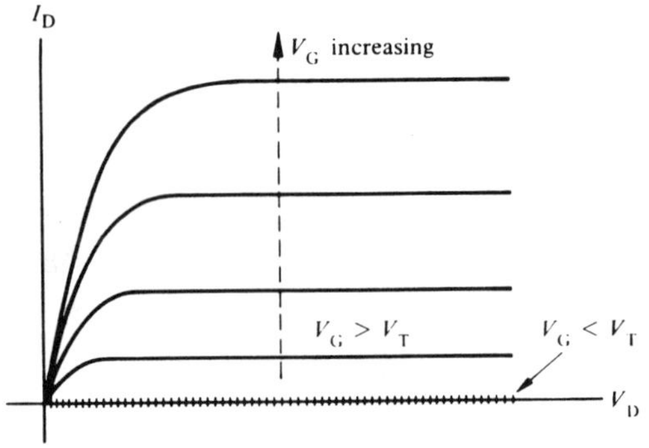

Figure 4. Drain current-voltage characteristics of a MOSFET device. From Pierret [8].

The MOSFET parameter of interest for chemical sensor application is V_T. Besides the metal-semiconductor work function difference, V_T, or equivalently V_G depends sensitively on charge distributions found in the oxide or at the interfaces.

An example of a chemical sensor based on a MOFSET is the Pd gate-hydrogen sensitive field effect device sometimes called a GASFET. Here Pd, which has a large hydrogen permeability, enables hydrogen atoms to diffuse to the metal-insulator interface where they form a dipole layer resulting in a shift in the I_D-V_G characteristics as illustrated in Fig. 5. For hydrogen in the presence of oxygen above 75° C

$$\Delta V \,\alpha \frac{(kP_{H_2}\ / Po_2)^{1/2}}{1+(kP_{H_2}\ / \ Po_2)^{1/2}} \tag{2}$$

Experimental results (see Fig. 6) show a shift in $\Delta V \simeq 0.4$V for 800ppm H_2 in Po_2=0.2 atm.

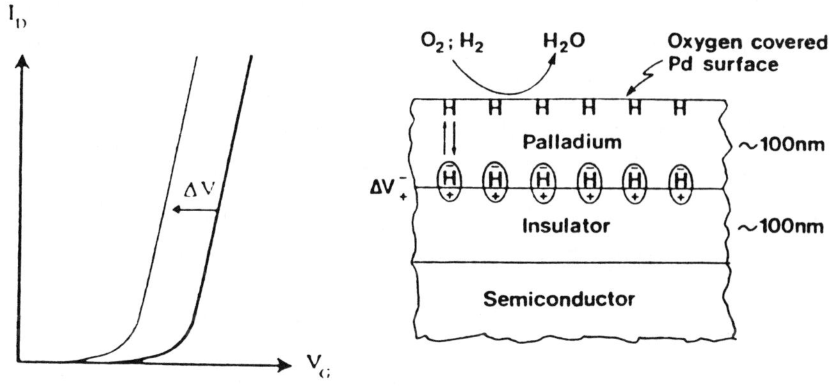

Figure 5. Shift in I-V characteristics of Pd gate MOSFET due to hydrogen coverage of Pd/SiO_2 interface. From Lundström [9].

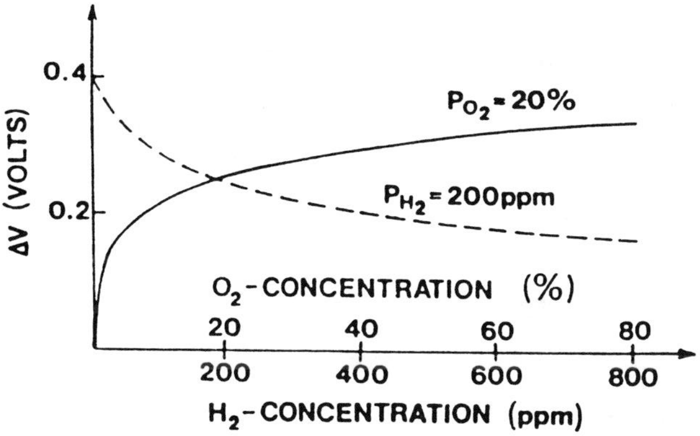

Figure 6. Response of Pd gate MOSFET due to P_{H_2} variation (P_{O_2}=0.2atm -solid curve) and P_{O_2} variation (P_{H_2} =200ppm -dashed curve) at 75°C. From Lundström [19].

MOFSET devices have not yet reached commercial use due to selectivity and response time issues. The hydrogen sensitive devices can also be sensitive to hydrocarbons, alcohols, hydrogen sulfide, etc. with the relative sensitivity depending on temperature (≤200° C) and the microstructure and catalytic nature of the gate. Response times typically range between 1-200 sec. Some devices, however, exhibit hysteresis and aging effects. Insulators other than SiO_2 appear to minimize the slow processes [9].

Integrated circuits contain many thousands of devices and it thus becomes possible to create sensors arrays in which devices include different catalytic metal gates (e.g. Pd, Pt, Ir) and/or operating temperatures. By use of pattern recognition, it becomes possible in this way to evaluate multicomponent gas mixtures [10].

SEMICONDUCTING GAS SENSORS

An alternative sensing mechanism relies on the dependence of the electronic conductivity of nonstoichiometric oxides on the oxygen partial pressure, P_{O_2} [6]. Reduction of $Gd_2Ti_2O_7$ is believed to occur by the reaction [11]

$$O_o \xrightarrow{Gd_2Ti_2O_7} V_O^{\bullet\bullet} + 2e' + \frac{1}{2}O_2 \qquad (3)$$

in which an oxygen ion is removed from the material to the gas phase leaving behind a doubly ionized oxygen vacancy and two conduction electrons. With the charge neutrality satisfied by

$$n = 2\left[V_O^{\bullet\bullet}\right] \qquad (4)$$

the electron density becomes

$$n = K(T)P_{O_2}^{-1/6} \quad (5)$$

where K(T) includes the mass action constant of reaction 3. More generally, the conductivity, σ takes the form

$$\log \sigma \ \alpha \ \pm \ 1/r \log P_{O_2} \quad (6)$$

where r is typically between 4 and 6, while (+) is for p-type and (-) is for n-type conduction [6]. The response of σ ($Gd_2Ti_2O_7$: Nb) to changes in P_{O_2} satisfies Eq (6) over broad limits [11].

At their lower temperatures, bulk diffusion becomes too slow and redox reactions become limited to exposed surfaces. Oxygen chemisorbed on the surface of an n-type semiconducting oxide will tend to deplete electrons from the near surface region. This may be detected by measuring a change in surface conductivity or in barrier height of a schottky barrier [12].

Bulk or thick film tin oxide-based gas sensors have been extensively used for, e.g., combustible gas alarms, cooking controls in microwaves and freon sensors [13]. The concern of gas selectivity, however, has always remained of major concern. Interesting developments along this line include the expanded use of sensor arrays with different catalysts or operating temperature enabling pattern recognition concepts to be applied, as was described in the case of the FET devices [10]. Instead of FET structures, here an array of thin film gas sensitive semiconducting oxide resistors are substituted. Another approach taken by Wu and Ko [14] takes advantage of the ability to rapidly cycle the temperature of thin film SnO_2 on a Si micromachined structure to obtain unique signatures for a series of hydrocarbon gases. More recently, Cavicchi et al. [15] have extended this idea to include an array of Si micromachined "microhotplates" which would allow for enhanced selectivity. A more detailed discussion of how electroceramics integrated with micromachined structures promise unique possibilities is presented in a later section.

SOLID ELECTROLYTE SENSORS AND BATTERIES

Electrochemical cells serve both as a source of electricity (batteries) and as the basis of sensors (concentration cell). Conventionally, the electrolyte of such cells is a liquid and is thus largely incompatible with silicon technology. The discovery of a number of key solid fast-ion conductors or solid electrolytes some 25 years ago [16, 17], initiated the development of the new field of "solid state ionics". While the lions share of activity in this field has focused on bulk materials, e.g., oxygen ion conduction in stabilized zirconia, sodium ion conduction in β-alumina, a growing number of studies have focused on thin films. This has been particularly true for electrochromic window applications where requirements of transparency and response time require thin films deposited over large planar arrays [18].

The key advantage of electrochemical over semiconductor sensors is that the output of the former is a thermodynamic quantity and can thus be used to measure specific activities of chemical species rather than relative changes. For example, the output of a lithium concentration cell with Li metal as an electrode is given by:

$$E = \frac{kT}{q} \ln a_{Li} \approx \frac{kT}{q} \ln C_{Li} \text{ [dilute solution)]} \qquad (7)$$

in which E is the cell emf, k the boltzmann constant, q the charge of the electron a_{Li} and C_{Li} the lithium activity and concentration respectively.

As in the case of the MOSFET and semiconductor sensors, integration of solid state ionic sensors into silicon chips opens up opportunities for monitoring complex gases, improved signal processing and lower cost. A recent example of such a device is a CO_2 sensor based on the sodium ion conductor $Na_4Zr_2(SiO_4)_3$ (NASICON) deposited as a thin film by RF magnetron sputtering on a silicon substrate [19]. As is the case for the other types of sensor, one key limitation which must be overcome is the relatively low operating temperatures that such devices are limited to.

Semiconducting CMOS-RAM memory microcircuits are central to the operation of most computers today. A key concern is the loss of memory during power outages. Thin film microbatteries, which could be integrated onto the chip, appear to be one attractive means for maintaining memory A schematic representation of a lithium microbattery is shown in Fig. 7. Issues relating to questions of fabrication, stability, rechargeability, storage capacity are recently reviewed by Julien and Nazri [21].

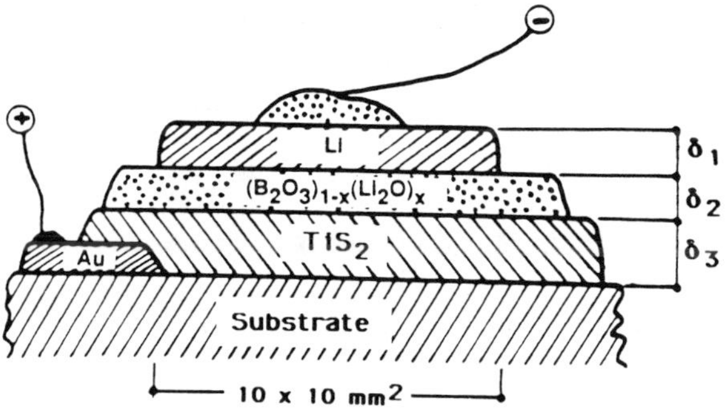

Figure 7. Schematic of a microbattery: Li/Li_2O-B_2O_3 glass electrolyte/TiS_2 insertion electrode. From Julien et al. [20].

An alternative to microbatteries for insuring that memory is not lost in semiconducting devices during power outages is the use of ferroelectric nonvolatile memories. Here, thin film capacitors are utilized in which the SiO_2 dielectric is replaced by a ferroelectric material such as lead zirconate titanate (PZT). Once polarized in a given direction towards the saturation polarization, removal of the field allows the polarization to relax to the remnant polarization. Since two opposite polarities are possible, this provides the basis for a nonvolatile binary memory. A schematic of a thin film ferroelectric capacitor on a CMOS device is illustrated in Fig. 8 [22].

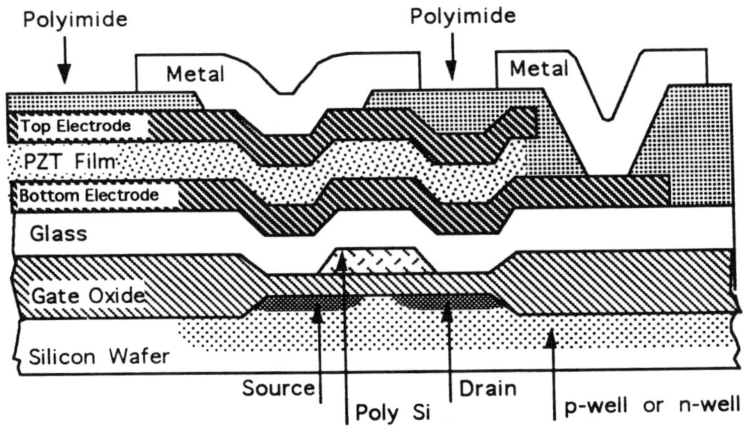

Figure 8. Schematic of a thin film ferroelectric capacitor on a CMOS device. From Blöm et al. [22].

To achieve useful devices, the ferroelectric films must satisfy a number of key criteria including low coercivity (low voltage operation), square hysteresis curves, non-fatiguing (i.e., ability to withstand 10^{12} read/erase/write cycles) and processing compatibility with CMOS technology. Sol gel and MOCVD processing approaches have provided good quality films at reduced temperatures. Efforts are presently focused on establishing the role of stoichiometry and point defects, degree of crystallinity, microstructure, and film-electrode interactions on remnant polarization, coercive fields and fatigue [23]. Progress continues to be made in this field. Ferroelectric random access memory (FRAM) devices appear to have a good possibility for commercial success.

MICROELECTROMECHANICAL DEVICES

Just as the introduction of microelectronics revolutionized the electronics industry some 30 years ago, the introduction of microelectromechanical (MEMS) devices promises to do the same for mechanical devices particularly those used as sensors or in some types of acutators. As suggested above, device fabrication and performance have been, in part, limited by the ability to micromachine the desired device geometries, and more specifically, to achieve micron or sub-micron sized, stress free, and high aspect ratio structures with precisely controlled dimensions.

We have been investigating the micromachining capabilities afforded by electrochemical (EL) and the photoelectrochemical (PEC) etching of silicon in HF electrolytes. This approach, we believe, holds promise for overcoming many of the limitations of conventional chemical micromachining [24, 25]. To investigate the suitability of EC and PEC etching for fabricating MEMS structures, a number of features were investigated including (a) etch rate and morphology dependence on cell bias, HF concentration, illumination intensity and wavelength and doping density, (b) development of high aspect ratio structures by photoassisted processes, and (c) development of p-n junction etch stops.

By appropriate choice of applied bias and HF concentration, the room temperature Si (p-type) etch rate was controllable over three orders of magnitude up to 100μm/min. This compares very favorably with the temperature-sensitive value of ~1μm/min achieved at approximately 80°C by, for example, KOH in the conventional chemical etching approach [26].

In contrast, n-type silicon which etches at a negligible rate in the absence of illumination, was found to increase its etch rate linearly with above band-gap illumination intensity. The feasibility of etching vertical high aspect ratio structures into n-Si by illumination through patterned metal masks was demonstrated [26].

To achieve selective etching of n- or p- regions, three novel and distinctly different p-n junction etch stops were developed [24-26]. Each etch stop was successfully demonstrated by micromachining cantilever beams with submicron thickness (~0.5μm) in which etch-rate selectivities as high as 10^8 were achieved. Furthermore, since these etch stops are compatible with low to moderate doping densities, this allowed for the fabrication of stress-free beams exhibiting virtually no curl over their 650μm length [27] (See Fig. 9).

Whichever processing approach is taken, the Si micromechanical structures serve as unique platforms onto which to deposit or integrate electroceramic materials. For example, recent successes in fabricating reproducible high quality ferroelectric films on silicon-based substrates for non-volatile random access memories has stimulated new work on the fabrication of micro-electro-mechanical sensors and actuators, and low mass pyroelectric thermal sensors [28]. Ferroelectric materials such as $Pb(Zr_xTi_{1-x})O_3$, or PZT and $PbTiO_3$, have high piezoelectric and pyroelectric coefficients making them particularly attractive for devices such as pressure sensors, pyroelectric infrared detectors, and flexure wave piezoelectric micromotors. Sayer et al. [28] estimate that the minimum detectable force F_m for a plate geometry transducer

$$F_m = \left(\frac{2kT\varepsilon\varepsilon_0 A}{\pi d_{33}^2 t} \right)^{1/2} \quad (8)$$

is equal to ≈ 7×10^{-7} N for a PZT device having thickness t=4μm, area A=0.1mm^2 (ε≈1000, d_{33}≈220×10^{-12}C/N). A cross section of such a device prepared by Polla et al. [29] is shown in fig. 10, in which the dimensions range from 20x20 μm^2 to 250x250 μm^2 with a diaphragm thickness made up of 0.45μm-thick PZT and 1.6μm- thick polycrystalline silicon. The measured response of the device was approximately 1μV/μbar of acoustic pressure at 1.0 kHz.

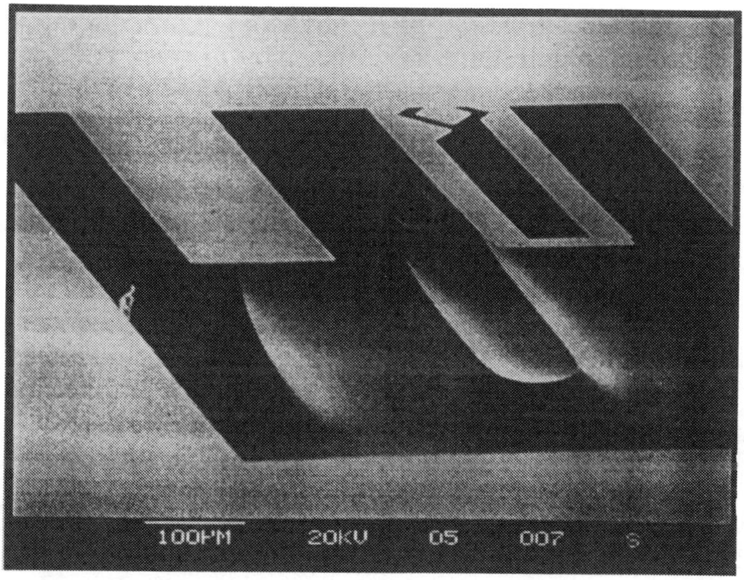

Figure 9. SEM micrograph of p-on-n cantilever beams (3.3 x 180 x 650μm) etched in an aqueous HF electolyte with the aid of illumination. From Mlcak [26].

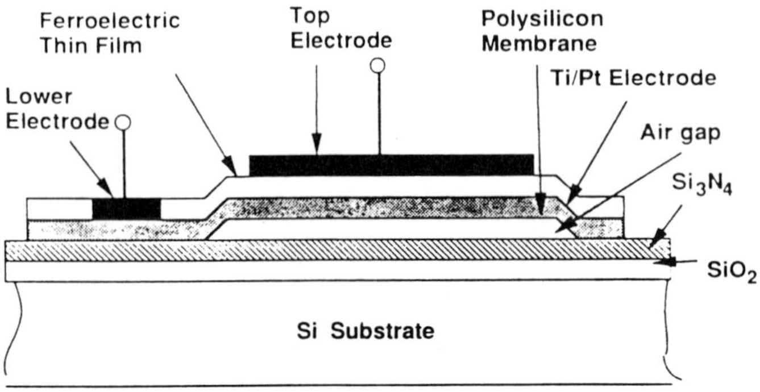

Figure 10. Schematic of a micromachined sensor with a ferroelectric thin film deposited onto a polycrystalline silicon membrane. From Polla et al. [29].

Pyroelectric sensors using similar structures benefit both from the low thermal mass of the membrane, the low thermal transfer surroundings and the relatively large pyroelectric coefficients. A device utilizing $PbTiO_3$ (p=90nC/cm^2-K, R.T.) achieved a voltage responsitivity of 4.0x10^4 V/W at a chopping frequency of 50Hz [29].

Micro-positioning actuators based on the inverse piezoelectric effect with linear displacements of up to 4μm [29] and macro-linear stepper-motors capable of sub-micrometer controlled movement [30] have been reported. Generally, such piezoelectric based devices show promise for low-noise, low-power consumption microsensors and large-force actuators [29]. Recently, even piezoelectric-on-silicon microrobots have been proposed [31]. Much attention is now being focused on achieving high quality ferroelectric films on the appropriate micromachined structures.

Semiconducting gas sensors, while very sensitive, often exhibit poor selectivity. However, Wu and Ko [32] recently demonstrated that SnO_2-based sensors show distinctly different signatures for closely related anesthetic agents if the temperature of the device can be cycled rapidly (~100°C/3min). This they were able to achieve by depositing SnO_2 films onto thin Si micromachined membranes (~3μm thick) with low thermal mass. A schematic of this device is shown in Fig. 11.

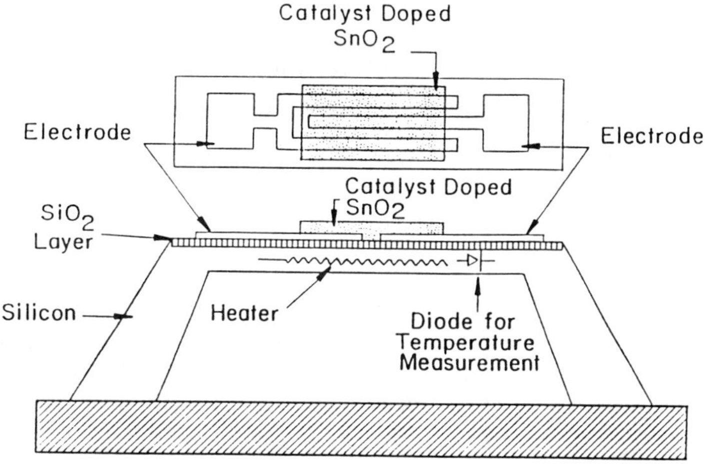

Figure 11. Schematic of a SnO_2 based semiconductor gas sensor integrated onto a micromachined Si membrane. From Wu and Ko [32].

SUMMARY

Electroceramics, as a class of materials, exhibit a wide versatility of properties. However, in many cases their use is being superseded due to their large dimensions, poor reproducibility or yield and/or high cost. Silicon based microelectronics, on the other hand, dominate the electronics industry due to their high device density and complexity achieved at low cost. However, silicon is limited in the ways it could be used for sensor or actuation functions, key to achieving closed-loop control of various processes. Integration of the two materials systems, particularly if one includes recent advances in MEMS technology, holds out great promise for creating sophisticated systems which include the necessary input and output functions as well as the signal processing capabilities of Si microchips. The challenge is largely centered on how to (1) retain the desirable bulk characteristics of electroceramics when applied in thin film form, (2) achieve this by processing methods compatible with Si microfabrication techniques, (3) insure materials compatibility-thermal, mechanical, and chemical, (4) and operate satisfactorily at temperatures typically below 200C. Much progress along these lines has already been made, particularly stimulated by FRAM development, and much more can be expected from the marriage of these two initially quite distinct materials technologies.

ACKNOWLEDGEMENTS

The author's research has been supported by the Basic Energy Sciences Division of the US Department of Energy (contract #DE-FG02-86ER45261) and the Center for Materials Science and Engineering at MIT (contract #DMR 90-22933).

REFERENCES

1. P.B. Vandiver and W.D. Kingery, in High Technology: Ceramics - Past, Present, and Future, ed. W.D. Kingery, American Ceramic Society, Westerville OH, 1986, p. 19.
2. R.H. Doremus, Glass Science, J. Wiley & Sons, New York NY, 1973, p. 2.
3. M. Berg, op. cit., ref. 1, p. 211.
4. L.E. Cross and R.E. Newnham, op. cit., ref. 1. p. 289.
5. A. Goldman, Modern Ferrite Technology, VanNostrand Reinhold, New York NY 1980, p. xi.
6. D.C. Hill and H.L. Tuller, in Ceramic Materials for Electronics, 2nd ed., R.C. Buchanan, Marcel Dekker, New York NY (1991), pp. 249-347.
7. J.S. MacDougall, *Application Bulletin*, Fairchild Semiconductor, November 1964.

8. R.F. Pierret, Field Effect Devices, 2nd ed., Addison-Wesley Publ. Co., Reading MA, 1990.
9. I. Lundström, in Fourth Int'l Meeting on Chemical Sensors, Tech. Digest, Japan Assoc. Chemical Sensors, Fukuoka, Japan, 1992.
10. U. Weimer, K.D. Shierbaum, W. Göpel and R. Kowalkowski, Sensors and Actuators B **1**, 93 (1990)
11. I. Kosacki and H.L.Tuller, "Donor-doped $Gd_2Ti_2O_7$ as a Semiconductor-Type Oxygen Sensor," Sensors and Actuators B **24-25,** 370-374 (1995).
12. M.H. Sukkar and H.L. Tuller, "Electronic Structure and Transport Properties of Interfaces in Metal Oxides," in Surface and Near-Surface Chemistry of Oxide Materials, Eds. J. Nowotny and L.-C. Dufour, Elsevier Sc. Publ., Amsterdam, 1988, pp. 611-67.
13. S. Matsuura, Op. cit., ref. 9, p.6
14. Q. Wu and W.H. Ko, Op. cit., ref. 10, p. 183
15. R.E. Cavicchi, J.S.Senhle, P. Chaparala, G.E. Poirier, K.G. Kreider, M. Gaitan, and S. Semancik, Fifth Int'l Meeting on Chemical Sensors-Technical Digest, vol. 2, Rome, Italy, 1994, p. 1136.
16. Fast Ion Transport in Solids, ed., W. van Gool, North Holland Pub. Co., Amsterdam, Holland, 1973.
17. H.L. Tuller and P.K. Moon, "Fast Ion Conductors - Future Trends," Mat. Sc. Eng. B, **1**, 171-91 (1988).
18. Electrochromic Materials II - Proc. of Symp., eds., K.-C. Ho and D.A. MacArthur, Electrochemical Society, Pennington NJ, 1994.
19. J.F. Currie, D. Ivanov, D. Ciureanu, A. Lecours, M. Caron, E. Sacher, G. Turcotte, and A. Yelon, in Ionic and Mixed Conducting Ceramics - Second Int'l Symp., Eds., T.A. Ramanarayanan, W.L. Worrell, and H.L. Tuller, Electrochemical Society, Pennington NJ, 1994, p. 127.
20. C. Julien, M. Massot, M. Balkanski, and H.L. Tuller, "Transport and Structure of Glasses for Microbattery Applications," in Glasses for Electronic Applications, (Ceramic Trans. Vol. 20) ed. K.M. Nair, American Ceramic Soc., Westerville OH, 1991, pgs. 51-84.
21. C. Julien and G.-A. Nazri, Solid State Batteries: Materials, Design, and Optimization, Kluwer Academic Publishers, Boston MA, 1994.
22. D.A. Blöm, T.A. Burr, and S.C. Hüguenin, 1994, unpublished.
23. G.H. Haertling, in Ferroelectric Films, Ceramic Trans., vol. 25, eds. AS. Bhalla and K. M. Nair, American Ceramic Society, Westerville OH, 1992, p. 1.
24. Electrochemical Etching Process, U.S. Patent No. 5,338,416, August 16, 1994 (with R. Mlcak)

25. P-N Junction Etch-Stop Technique for Electrochemical Etching of Semiconductors, U.S. Patent No. 5,464,509, November 7, 1995 (with R. Mlcak)
26. R. Mlcak, Electrochemical and Photoelectrochemical Micromachining of Silicon in HF Electrolytes, PhD thesis, Dept. of Materials Science and Engineering, MIT, Cambridge MA, 1994.
27. R. Mlcak, H.L. Tuller, P. Greiff, J. Sohn, and L. Niles, "Photoassisted Electrochemical Machining of Silicon in HF Electrolytes," Sensors and Actuators A, **40**, 49-55 (1994).
28. M. Sayer, C.V.R. Vasant Kumar, D. Barrow, L. Zou, and D.T. Amm, in Ferroelectric Thin Films II (MRS Proc. Vol. 243), eds. A.J. Kingon, E.R. Meyers, and B. Tuttle, Materials Research Society, Pittsburgh PA, 1992, p. 39.
29. D.L. Polla, C. Ye, P. Schiller, T. Tamagawa, W.P. Robbins, D. Glumac, and C.C. Hseuh, ibid., p. 55.
30. W.P. Robbins, IEEE Trans. on Ultrasonics, Ferroelectrics, and Frequency Control, **38** 461 (1991), Quoted in ref. 29.
31. J.G. Smits, Sensors and Actuators A **35** 129 (1992)
32. Q. Wu and W.H. Ko, Sensors and Actuators B **1** 183 (1990)

SUBSTRATES FOR MICROWAVE APPLICATIONS OF HIGH Tc SUPERCONDUCTORS

A.S. Bhalla and Ruyan Guo
Materials Research Laboratory, The Pennsylvania State University
University Park, PA 16802

ABSTRACT

Systematic studies on low κ oxide substrate materials suitable for epitaxial deposition of high Tc superconducting (HTSC) yttrium barium cuprate (YBCO) films and their microwave applications have been carried out. Several promising new hosts such as $Sr(Al_{1/2}Ta_{1/2})O_3$, $Sr(Al_{1/2}Nb_{1/2})O_3$, and $Ba(Mg_{1/3}Ta_{2/3})O_3$ in complex oxide perovskite family have been developed. The crystals of these compositions and their associated solid solutions with $LaAlO_3$ and $NdGaO_3$ provide new options for twin free, ultra low loss, low permittivity substrates with close structural and thermal properties matching to the YBCO. The unique capabilities of a laser heated pedestal growth (LHPG) system have been utilized for test-growth of candidate materials in single crystal fiber form to determine structure, thermal, and dielectric properties and to make positive identification of twin free systems.

INTRODUCTION

Superconductor device applications demand that the substrate materials for HTSC thin films meet the following requirements in general:

- twin-free single crystal with matching lattice (or matching atomic structure) to the HTSC for oriented or ideal epitaxial films;
- thermal compatibility in terms of thermal expansion matching over the temperature range of film processing and operating (90K to >650°C), and no structural phase transitions in this temperature range;

To the extent authorized under the laws of the United States of America, all copyright interests in this publication are the property of The American Ceramic Society. Any duplication, reproduction, or republication of this publication or any part thereof, without the express written consent of The American Ceramic Society or fee paid to the Copyright Clearance Center, is prohibited.

- suitable dielectric properties (dielectric constants and loss tangent) at liquid nitrogen temperature and microwave frequency;
- chemically inert at deposition and O_2-annealing temperatures; high stability in the ambient over time; and
- no cleavage property in the surface plane of interface; mechanically strong and scratch resistant.

In addition to satisfy the general requirements, substrates to be used in microwave devices must be twin-free single crystals with potential to be grown in large size (2-4" wafer). Congruent melting compositions with modest melting temperatures are therefore favorable. The dielectric constant influences the propagation speed and the package dimensions therefore a moderately low dielectric constant (~25) is required. Because the conductor loss of the superconductor is diminished below T_c, the role of the dielectric loss becomes critically important. Very low dielectric loss ($<10^{4}$) is therefore demanded.

The potential application of YBCO superconductor to the development of ultra high density interconnect systems for a new generation of high speed high density multichip modules dictates new requirements for the substrate design. An essential feature is that the X and Y layers of HTSC lines must have geometry in cross section of the order 2 x 1 μm thus necessitating a current carrying capability for the HTSC of $\sim 10^6$ A/cm^2. For this current capability clearly the YBCO must be highly grain oriented very near to single crystal, so that the whole multilayer structure is in the form of highly oriented overgrowths on a single crystal substrate. Impedance characteristics for the interconnect structure dictate that the dielectrics used must be of low permittivity (ideally <10) if the geometry is to be preserved. Highly oriented thin films (2 μm or above) will be needed for the separation of strip line and ground plane structures. Dielectric loss requirements are somewhat less stringent than those required for the microwave applications; however, $\tan\delta < 0.001$ is highly desirable at the 77K working temperature.

Substrates with the perovskite structure are among the most obvious candidates for epitaxial growth of YBCO films. Those substrates of perovskite structure usually have ideal cubic (such as $SrTiO_3$), double cell cubic (such as Ba_2YSbO_6), hexagonal distorted pseudo cubic (such as $LaAlO_3$) or other pseudo cubic cells (such as $NdGaO_3$ that has the [$GdFeO_3$] structure with orthorhombic symmetry). $SrTiO_3$ crystals of high quality and large sizes, though are readily available and yield the best epitaxial quality thin films (primarily due to their close interatomic structure matching to the HTSC), have high dielectric constant (>300) arising from the phase transition near the working temperature (~110K). Another probably the most used substrate currently, $LaAlO_3$, though good in lattice matching and of reasonable dielectric properties, is heavily twinned and goes through a ferroelastic phase transition at ~435°C.[1] These macroscopic defects in the crystals are not tolerable in substrates used for complex microcircuits. Table I lists some of the most commonly used substrates for comparison purposes.

Table I. Most commonly used high Tc superconductor substrates

Substrates	Symmetry, Lattice Constants (Å)	Dielectric Constant κ (at RT)	Thermal Expansion α (x10^{-6}/°C)	Comments
$YBa_2Cu_3O_{7-\delta}$	Ortho. a=3.836 b=3.883 c=11.68 (Ref. [2])		α_a=14 α_b=9 α_c=19 (Ref. [3])	
$SrTiO_3$	Cubic a=3.905	>300	10.6	Good epitaxial YBCO film; high dielectric constant
$LaAlO_3$	Rhomb. a=3.789 α=90.12°	23	~10	Structural phase transition at ~435°C; heavy twinning
Al_2O_3	Rhomb. a,b=4.7586 c=12.9897 γ=120°	κ_a=9.34 κ_c=11.54	6	Poor lattice matching; poor thermal expansion matching; chemical interactions with YBCO (buffer layer often required)
MgO	Cubic a=4.2114	9.65	10.4	Reasonable cost; easy to cleave; large lattice mismatching
YSZ ($Zr_{0.72}Y_{0.28}O_{1.862}$)	Cubic a=5.16(2)	26	4.5	Chemical interactions with YBCO, $BaZrO_3$ forms at the interface (buffer layer often required)

In this paper, complex oxide perovskite substrate candidates such as $Sr(Al_{1/2}Ta_{1/2})O_3$ and $Sr(Al_{1/2}Nb_{1/2})O_3$, $Ba(Mg_{1/3}Ta_{2/3})O_3$ in complex oxide perovskite structure family and their solid solutions with ternary perovskites $LaAlO_3$ and $NdGaO_3$ are summarized.

CRYSTALS OF COMPLEX OXIDE PEROVSKITE COMPOUNDS

Ceramic samples were prepared by solid state reaction, using conventional techniques. X-ray diffraction technique was used extensively to characterize the crystallographic phases and to adjust the processing conditions. The laser heated pedestal growth (LHPG) method has been shown to be a powerful method for rapidly growing small diameter single crystals, particularly oxides of high melting temperature, for both property studies and fiber devices.[4,5,6] The LHPG equipment used in this investigation consisted of a power source (water cooled, tunable flowing gas CO_2 55W laser), an optical layout, and a growth section. The molten zone temperature during a stable growth was monitored using an optical pyrometer.

Radio frequency dielectric constants and the loss tangent were measured using a General Radio 1621 Capacitance Measurement System. Dielectric properties at microwave frequency were measured using resonance techniques equipped with an HP8510A network analyzer. Post resonance technique (the Hakki and Coleman technique) was used to measure the dielectric constants of the ceramic samples. Cavity perturbation technique was used for the measurements on samples of thin rod (e.g., single crystal fiber samples) or bar-shaped. The Q factors (of microwave frequency) at liquid nitrogen temperature were measured by a transmission resonance technique.

$Ba(Mg_{1/3}Ta_{2/3})O_3$ (BMT)

Ceramics of complex perovskite oxides $A(B1_{1/3}B2_{2/3})O_3$ type have been explored previously as microwave dielectric materials.[7] $Ba(Mg_{1/3}Ta_{2/3})O_3$ (BMT), in particular, was reported to have a dielectric constant $\kappa \sim 25$ and dielectric Q~16,800 (one of the highest in the oxide family) at 10.5 GHz, in samples with 1 mol% Mn additive as a sintering aid.[8] BMT compound is one of the most refractory oxides known thus the growth of single crystals is difficult. A single crystal of BMT was grown from a BaF_2 flux. It yielded a significantly higher dielectric constant (~200),[9] presumably attributable to the flux contamination.

BMT single crystal fibers were grown successfully using our LHPG technique. It grows congruently from the melt in the temperature range of 2900-3100°C. A high temperature phase of simple cubic perovskite was obtained at room temperature, in comparison to the hexagonal ordered perovskite structure usually obtained in ceramics (see Fig. 1). Dielectric properties of both the ceramic and the single crystal BMT were studied. BMT ceramic samples have ultra low dielectric loss ($<1x10^{-5}$ at 90K and 10kHz) and good thermal compatibility ($\alpha \sim 9.0x10^{-6}/°C$) with the YBCO superconductors. The single crystal BMT has a cubic lattice parameter a=4.0877Å. The dielectric constant increases and saturates as the bulk density approaches the theoretical density. Dielectric loss decreases with the enhancement of the ordering of the B-site. Single crystals of high temperature disordered cubic form preserve a moderate dielectric constant (26.0 at 10GHz) and low dielectric loss tanδ ($2.78x10^{-4}$ at room temperature and 10kHz and $<10^{-5}$ at 90K) that make this material unique for microwave device applications.

The BMT lattice parameter of a=4.0877Å, represents a lattice mismatch of 5.3% to the b-axis of YBCO (b=3.883Å); this seems less ideal as a substrate for YBCO. However, there has been no clear cut-off for lattice parameter matches for "epitaxial" (or highly oriented) film deposition of YBCO. "Epitaxial" YBCO thin films on MgO single crystals (with mismatch of 8.5%) have been reported.[10] A BMT single crystal has a twin-free cubic perovskite structure that is advantageous as a substrate compared to some of the heavily twinned substrates, e.g., $LaAlO_3$ and $NdAlO_3$. High temperature BMT single crystal grown by LHPG is twin free, of moderate dielectric constant, low dielectric loss, and good thermal expansion matching and is therefore identified to be a potentially suitable substrate for the HTSC thin film deposition.

The application of BMT as a substrate, beside its fiber crystals' applications for microwave antenna, may be restricted by the fact that single crystals are difficult to grow. Skull melting growth techniques,[11] could presumably be used to grow BMT crystals of adequate sizes. The high melting temperature of BMT will not be a crucial issue, when the material is used as an insulating layer between the YBCO films in a multichip-module type of integrated structure, because vapor phase deposition techniques (e.g., laser ablation and metal-organic chemical vapor deposition) rather than liquid phase growth methods will be utilized.

$Sr(Al_{1/2}Ta_{1/2})O_3$ (SAT) and $Sr(Al_{1/2}Nb_{1/2})O_3$ (SAN)

The compounds of $Sr(Al_{1/2}Ta_{1/2})O_3$ (SAT) and $Sr(Al_{1/2}Nb_{1/2})O_3$ (SAN) were first prepared and tested to learn their crystallographic phases and melting behavior by the group at the AT & T Bell Labs.[12] Ceramic samples were identified to have double cell cubic perovskite structure with a=7.795Å and melting temperatures of 1900°C and 1790°C for SAT and SAN, respectively. On the basis of our understandings of the crystal chemistry-dielectric property relation of various oxide perovskites, and the reports by the Bell Labs group that showed both the SAN and SAT melt congruently and produce a single phase of the perovskite structure after melting, we selected the SAT and SAN as primary candidates in the

$A(B1_{1/2}B2_{1/2})O_3$ complex oxide perovskite family for crystal growth and to investigate their properties in relation to substrate applications.

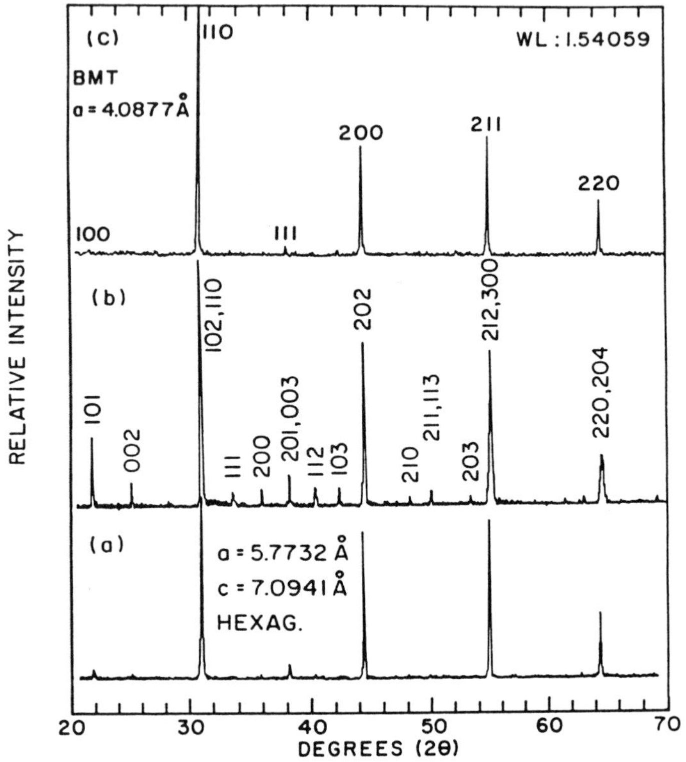

Fig. 1. Typical x-ray diffraction pattern of $Ba(Mg_{1/3}Ta_{2/3})O_3$: (a) ceramic powder calcined at 1500 °C for 3 h showing weak ordering; (b) ceramic sintered at 1655 °C for 5 h showing strong B-site ordering; and (c) single crystal (powder) grown by LHPG showing an ideal simple cubic perovskite structure with a=4.0877 Å. BMT ceramic XRD patterns are indexed as a hexagonal perovskite structure with *a*=5.7731(6) Å and *c*=7.0941(2) Å.

$Sr(Al_{1/2}Ta_{1/2})O_3$ (SAT) and $Sr(Al_{1/2}Nb_{1/2})O_3$ (SAN) were grown using the LHPG growth technique. Their crystallographic structures were found to be simple cubic perovskite with lattice parameters a=3.8952Å (SAT) and a=3.8995Å (SAN) that are of close lattice matching to the YBCO superconductors. No structural phase transitions or twins have been found and the average coefficients of the thermal expansion are in good matching with the YBCO superconductor materials. SAT currently represents one of the best potential HTSC substrate materials for microwave applications. The unique feature of this material is that it has desired dielectric properties (dielectric κ~12, loss factor $\tan\delta<10^{-4}$) at the microwave frequencies with twin-free lattice and good thermal expansion matching along with chemical compatibility with the YBCO superconductors. Dielectric constants in RF region as measured on the dense ceramic samples of SAT and SAN at 23°C are shown in Fig. 2.

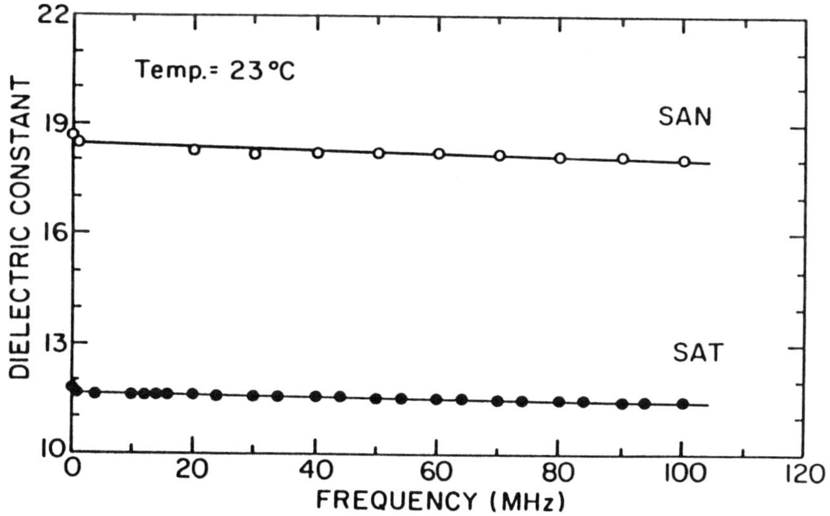

Fig. 2. Dielectric constants of SAT and SAN ceramics as functions of frequency at room temperature.

Our experimental results on the SAT composition along with the earlier report on the congruent melting nature of the SAT and SAN compounds, have stimulated research works in the thin film area. MOCVD-derived SAT films grown at 850°C on $LaAlO_3$ were found to have exclusively (001) growth with in-plane orientation.[13] Using SAT polycrystalline materials as targets in a pulsed laser deposition process, Findikoglu *et al.*[14,15] have reported high quality epitaxial (c-axis orientation) growth of thin films of SAT and multilayers of YBCO/SAT on (001) $LaAlO_3$ and MgO substrates. Dielectric constants of the SAT films (~100-390nm thick), however, were reported to be ~23-30, notably higher than the value (κ~12) found in bulk ceramic materials. The deviations from perfect cation stoichiometry in films may be one of the causes for the discrepancy in value of the dielectric constants as the Al/Ta ratio in SAT films was found to be ~0.8 rather than the stoichiometric ratio of unity.[12,13] No dielectric loss data has been reported for the SAT films therefore direct comparison between the dielectric constants of the film and that of the bulk SAT sample is not intended.

SOLID SOLUTION OF TERNARY AND COMPLEX PEROVSKITE OXIDES

Further modification of the SAN and SAT compositions has been carried out to fine tune their properties, particularly reducing their melting temperatures (SAT: 1900~(1908±25)[12], and SAN:1790~(1739±10)[12])[16] for easier fabrication of the crystals and better control of the reduction problem of the Nb^{5+} and Ta^{5+}. This modification was also an attempt to overcome the twinning problem and to stabilize the cubic phase at room temperature in $LaAlO_3$ crystals.

$LaAlO_3$ has a rhombohedrally distorted perovskite ($A^{3+}B^{3+}O_3$ type) structure. Although the La^{3+} ion generally prefers the 12-coordination-site, it has a tendency for 9-coordination. The distortion in the LaO_{12} polyhedron is brought about by a slight displacement of the oxygen atoms away from the ideal positions of the cubic perovskite form, that is shown more clearly in other [rear-earth]$^{3+}AlO_3$ family members when the A-site cation radii become even smaller, e.g., in the case of $PrAlO_3$.[17,18] Fig. 3 presents a classification of $[A]^{3+}[B]^{3+}O_3$ type compounds according to the constituent ionic radii (8-coordination cation radii were used for A-site cations) focusing on the perovskite region. In fact, no ideal cubic perovskite structure but the rhombohedral [$LaAlO_3$] and the orthorhombic [$GdFeO_3$] structure have been reported in ternary compounds of the $[A]^{3+}[B]^{3+}O_3$ type.[19] For aluminate compounds, rhombohedral symmetry is found when A-site is occupied by the largest A^{3+} cation, La^{3+}, and other $[A]^{3+}[Al]^{3+}O_3$ compounds have even lower symmetry.

Our approach following this direction was to introduce "balanced" cation substitution simultaneously in the A- and B-sites to increase the average cation size at the A-site, hence to stabilize the 12-coordination of that position and consequently the cubic perovskite structure. The solid solution of ternary $LaAlO_3$ and complex oxides of $Sr(Al_{1/2}Ta_{1/2})O_3$ or $Sr(Al_{1/2}Nb_{1/2})O_3$ was chosen for investigation.

By forming crystalline solutions with compounds of low melting temperatures, it was expected that the crystalline solution would result in lower melting temperature and consequently avoid the reduction problem and permit growth in platinum crucibles. $NdGaO_3$ was found to have a melting temperature of ~1484±

24°C, and it was therefore selected as an end member of the crystalline solution series with SAT and/or SAN for the present studies. $NdGaO_3$ has the [$GdFeO_3$] structure with orthorhombic symmetry. No twinning problems are reported in this material. YBCO thin films deposited on $NdGaO_3$ were of better quality compared to those deposited on $LaAlO_3$ substrates. However, the high dielectric loss in the $NdGaO_3$ is a limiting factor for the YBCO film applications in microwave devices.

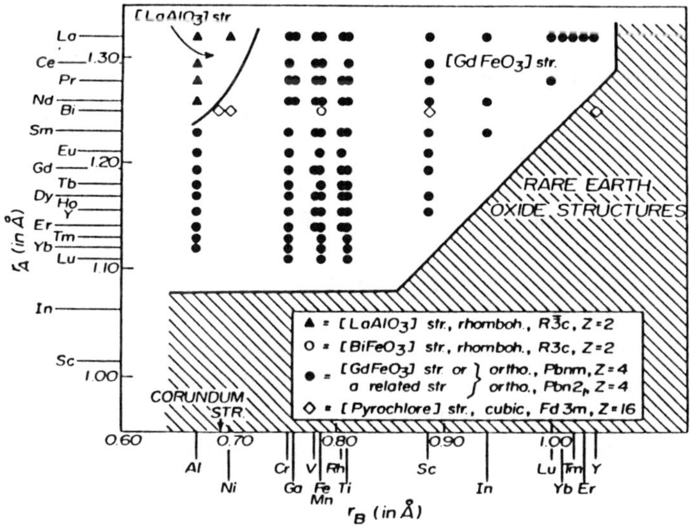

Fig. 3. Structure field map for $[A]^{3+}[B]^{3+}O_3$ type compounds (ref. 19).

Single crystal fibers of modified strontium aluminum tantalum oxide $(1\text{-}x)Sr(Al_{1/2}Ta_{1/2})O_3{:}xLaAlO_3$ (SAT:LA) and $(1\text{-}x)Sr(Al_{1/2}Ta_{1/2})O_3{:}xNdGaO_3$ (SAT:NG), and modified strontium aluminum niobium oxide $(1\text{-}x)Sr(Al_{1/2}Nb_{1/2})O_3{:}xNdGaO_3$ (SAN:NG) and $(1\text{-}x)Sr(Al_{1/2}Nb_{1/2})O_3{:}xLaAlO_3$ (SAN:LA) were grown using a laser heated molten zone growth technique.[20] 0.7SAT:0.3LA grows congruently and remains twin free simple cubic perovskite structure (as the SAT) when cooled down to room temperature. 0.7SAT:0.3LA crystals have moderate dielectric constant (κ=21.7) and low dielectric loss ($\tan\delta = 7.5 \times 10^{-5}$) at 10 kHz and 90K. The reduction problem of Ta^{5+} is eliminated (which is common in the case of SAT growth). 0.7SAT:0.3NG and 0.7SAN:0.3NG have lower melting temperatures and crystal growth is easier. $NdGaO_3$ addition to the SAT and SAN enhances the potential of SAT and SAN as large area substrates for HTSC growth. However, the dielectric constants increased from κ~12 to κ~16 (0.7SAT:0.3NG) and from κ~18 to κ~23 (0.7SAN:0.3NG) as a result of $NdGaO_3$ incorporation.

Our results further confirmed the report of Mateika *et al.*[21,22] that the ideal cubic phase can be formed in $(La,Sr)(Al,Ta)O_3$ compounds. It is interesting to notice that similar substitutions using [Ca,Ta] instead of [Sr,Ta] did not produce a compound with cubic structure. The average A-site cation radii of the [Ca,Ta] substitution is smaller than that of $LaAlO_3$ (ionic radii of Ca^{2+}, La^{3+}, and Sr^{2+} are 1.14, 1.185, and 1.27Å, respectively),[23] therefore, no stabilization effect on 12-coordination A-site is expected.

The cubic symmetry for the compound of SAT-NG and SAN-NG may be due to the fact that Ga^{3+} is almost of the same cation size as Ta^{5+}/Nb^{5+}. Slight reduction in the A-site cation size is accompanied by the slight increase of the B-site cation size and thus the cubic structure of SAN or SAT stays intact.

The ideal cubic perovskite structure can be stabilized in the case of ternary $LaAlO_3$ by forming a crystalline solution composition with cubic $Sr(Al_{1/2}Ta_{1/2})O_3$ and $Sr(Al_{1/2}Nb_{1/2})O_3$. The mechanism of this type of stabilization is through introducing the compensated cation substitution in the form of $[2La^{3+}]+[Al^{3+}] \rightarrow [2A^{2+}]+[B^{5+}]$ with the A^{2+} cation having a radius larger than that of La^{3+} and therefore stabilizing the 12-coordinated A-site. Crystalline solutions of SAT-LA maintained or improved most of the dielectric and thermal properties of $LaAlO_3$ and gained the advantage of forming a twin-free simple cubic structure and improved lattice compatibility. $NdGaO_3$ is shown to be an effective end member to decrease the melting temperature of SAN and SAT without disturbing their simple cubic (twin-free) crystal structure. Dielectric constants of SAN and SAT with addition of the $NdGaO_3$ were increased; however, the dielectric loss factor still remained less than $5x10^{-4}$. The results suggesting that SAT-LA and SAN-LA are better candidates as substrate materials than $LaAlO_3$ because the latter is intrinsically twinned. The growth of SAT-NG and SAN-NG are comparatively convenient as they have relatively low melting temperatures together with the relatively lower dielectric constants and the ideal lattice constants and thermal compatibility with the YBCO superconducting materials.

Other substrate candidates such as $La(Mg_{2/3}Ta_{1/3})O_3$,[24] $La(Mg_{1/2}Ti_{1/2})O_3$,[25] $(Ca,Sr)(Ga,Nb)O_3$[26] and a family of materials of the magnetoplumbite structures[27] have also been fabricated and their dielectric properties studied. These crystals and their associated solid solutions provide several new options for ultra low loss, low permittivity, twin free oxides with low congruent melting temperature, matching thermal expansion and excellent chemical compatibility.

SUMMARY

Oxide crystals with the perovskite structure are major candidates for YBCO film epitaxial deposition particularly when large size, cubic symmetry, and twin free crystals become available. SAT has already shown promise in high quality epitaxial YBCO film growth. With improved growth capability, the SAT and particularly its modified solid solutions may be more useful than the intrinsically twinned $LaAlO_3$. Solid solutions between the known complex perovskite oxide and ternary end members, especially gallates, will provide large number of options to tailor the material for specific device applications. The property parameters of the newly proposed substrate materials of oxide perovskite family are summarized in Table II. $LaAlO_3$ and $NdGaO_3$ are also listed for comparison.

TABLE II. Properties of Some Newly Proposed Oxide Perovskite Substrate Materials

Composition	Symmetry	Lattice Constant (Å)	κ 10kHz 90K	tanδ 10kHz 90K	α ($x10^{-6}$/°C) (@~300K)	Melting Point (°C) *
BMT (crystal)	Cubic	4.0877	25.9	$<10^{-5}$	9.0	(>2800)
SAT	Cubic	3.8952	11.8	$4.2x10^{-5}$	9.7	1908±25
SAN	Cubic	3.8995	18.7	$2.2x10^{-4}$	8.5	1739±10
SAT-LA	Cubic	3.8727	21.7	$7.5x10^{-5}$	7.7	1830±22
SAN-LA	Cubic	3.8634	25.7	$2.8x10^{-4}$	9.5	1705+20
SAT-NG	Cubic	3.8866	16.0	$4.3x10^{-4}$	8.8	1767±31
SAN-NG	Cubic	3.8790	23.0	$5.2x10^{-4}$	10.8	1582±20
$LaAlO_3$ (crystal)	Rhomb.	a=3.789 a=90.12°	23	$7.5x10^{-5}$	8.2//$[1\bar{1}0]$ 6.4//[001]	2040±9
$NdGaO_3$ (crystal)	Ortho.	a=5.426 b=5.502 c=7.706	23 (77K,	$3.2x10^{-4}$ 10GHz)		1484±24

*The melting point was determined rather simply using a strip furnace, with two operators using two separate optical pyrometers and averaging several readings per sample.

ACKNOWLEDGMENT

This work was supported by the Defense Advanced Research Projects Agency (DARPA) under the contract No. DN 00014-90-J-4140.

REFERENCES

1 E.A. Wood, *Amer. Min.* **36** 768 (1951).

2 W. Wong-Ng, F.W. Gayle, D.L. Kaiser, S.F. Warkins, and F.R. Fronczek, *Phys. Rev. B* **41**(7), 4220 (1990).

3 Hoydoo You, U. Welp, and Y. Fang, *Phys. Rev.* **B43** (4), 3660 (1991).

4 J.S. Haggerty, W.P. Menashi and J.F. Wenckus, U.S. Patent No. 3 944 640 (16 March 1976); U.S. Patent No. 4 012 213 (15 March 1977).

5 R.S. Feigelson, *MRS Bull.* **13** 47 (1988).

6 J. Yamamoto and A.S. Bhalla, *Mat. Res. Bull.* **24** 761 (1989).

7 K. Wakino, *Ferroelectrics* **91** 69 (1989).

8 S. Nomura, K. Toyama, and K. Kaneta, *Jpn. J. Appl. Phys.* **21**(10) L624 (1982).

9 F. Galasso and J. Pinto, *Nature* vol. **207** No. 4992, 70 (1965).

10 e.g., T. Terashima, K. Iijima, K. Yamamoto, K. Irata, Y. Bando, and T. Takada, *Jpn. J. Appl. Phys.* **28**, L987 (1989).

11 H.R. Harrison and J.M. Honig, *Bull. Mater. Sci.* **3**(3), 247 (1981).

12 C.D. Brandle and V.J. Fratello, *J. Mater. Res.* **5**(10), 2160 (1990).

13 B. Han, D. A. Neumayer, B.H. Goodreau, T. J. Marks, H. Zhang, and V.P. Dravid, *Chem. Mater.* **6**, 18 (1994).

14 A.T. Findikoglu, C. Doughty, S. Bhattacharya, Qi Li, X.X. Xi, T. Venkatesan, R.E. Fahey, A.J. Strauss, and J. M. Phillips, *Appl. Phys. Lett.* **61** 1718 (1992).

15 A.T. Findikoglu, S. Bhattacharya, C. Doughty, M.S. Pambianchi, Qi Li, X.X. Xi, S.M. Anlage, R.E. Fahey, A.J. Strauss, J.M. Phillips, and T. Venkatesan, *IEEE Trans. Appl. Superconductivity*, **3**(1) 1425 (1993).

16 R. Guo, A.S. Bhalla, J. Sheen, F. Ainger, E.C. Subbarao, S. Erdei, and L.E. Cross, *J. Mat. Res.* **10**(1), 18 (1995).

17 R.D. Burbank, *J. Appl. Cryst.* **3** 112 (1970).

18 O. Muller and R. Roy, in "*The major ternary structural families*" (Springer-Verlag, Berlin, Heidelberg, New York 1974) p.215.

19 F.S. Galasso, in "*Structure, properties and preparation of perovskite-type compounds*" (Pergamon Press, Oxford, London, Edinburgh, Now York, Toronto, Sydney, Paris, Braunschweig 1969) p.10.

20 R. Guo, P. Ravindranathan, U. Selvaraj, A.S. Bhalla, L.E. Cross, and R. Roy, *J. Mater. Sci.* **29**(19), 5054 (1994).

21 S. Haussühl and D. Mateika, *Crystal. Res. Technol.* **26**(4) 481 (1991).

22 D. Mateika, H. Kohler, H. Laudan and E. Völkel, *J. Cryst. Growth* **109** 447 (1991).

23 R.D. Shannon and C.T. Prewitt, *Acta Cryst.* **B25** (1969) 925; *Acta Cryst.* **B26** 1046 (1970).

24 P. Ravindranathan and A.S. Bhalla *et al.* (to be published).

25 G. Harshe, A.S. Bhalla, and L.E. Cross, *Materials Letters* **18**(4), 173 (1994).

26 S. Erdei, L.E. Cross, F.W. Ainger, and A.S. Bhalla, *J. Cryst. Growth* **139**, 54 (1994).

27 R. Guo and A.S. Bhalla (to be published).

PROCESSING AND PROPERTIES OF BARIUM TITANATE/POLYMER HYBRID MATERIALS BY SOL-GEL METHOD

Shin-ichi Hirano, Toshinobu Yogo, Ko-ichi Kikuta and Seiji Yamada
Department of Applied Chemistry, School of Engineering, Nagoya University, Furo-cho, Chikusa-ku, Nagoya 464-01, Japan

ABSTRACT

Barium titanate ($BaTiO_3$)/polymer nano-composite was successfully synthesized by methacryltriisopropoxytitanium (MTPT), barium alkoxide and styrene. $BaTiO_3$ particles/polymer hybrid material was formed after the polymerization and hydrolysis of the complex alkoxide. Crystalline $BaTiO_3$ particles of around 3 nm in size were dispersed through in-situ formation in the polymer matrix. The $BaTiO_3$/styrene hybrid films were synthesized from Ba-Ti complex alkoxide and styrene. The dielectric properties of the hybrid films were measured.

INTRODUCTION

Organically modified ceramics, such as Ormocer, Ormosil and Ceramer [1-3], have been receiving great attentions because of their applications in newly developed composite materials. Silicate and silica based systems are modified based upon the different reactivity between Si-O and Si-C bonds toward hydrolysis. Inorganic backbone is constructed by the condensation of Si-O bonds, whereas organic chain is formed by radical polymerization of C=C bonds. On the other hand, the control of reaction of transition metal-organics is more difficult compared with the silica based systems, since the transition metal-organic bonds (M-OR, M-C) are more susceptible to hydrolysis than Si-C bonds. The transition metal oxide/organic materials, therefore, are limited to the TiO_2 and ZrO_2 based systems [4,5]. The novel properties of the inorganic/organic hybrid materials originate from the nano-scale structure including bond natures between organic phase and inorganic

To the extent authorized under the laws of the United States of America, all copyright interests in this publication are the property of The American Ceramic Society. Any duplication, reproduction, or republication of this publication or any part thereof, without the express written consent of The American Ceramic Society or fee paid to the Copyright Clearance Center, is prohibited.

phase.

The dielectric properties of barium titanate depend upon the size and crystal regularity of grains. Usually, crystalline $BaTiO_3$ particles above 0.1μm were dispersed in organic media using a conventional mixing method in order to fabricate dielectric/organics composites. The nucleation and crystallization in an organic matrix through sol-gel process allow to obtain $BaTiO_3$ particles of much smaller size bonded to the organic matrix. The authors reported the synthesis of crystalline $BaTiO_3$ particles in a polymer matrix using Ba-Ti complex alkoxide [6]. Ultra-fine ferroelectric particles in a polymer matrix is expected to behave like superparaelectrics. The control of $BaTiO_3$ in particle size, crystallinity, and bonding to the matrix are key parameters for the design of the new composite materials.

This paper describes the synthesis of barium titanate/polymer hybrid from barium alkoxide, methacrylisopropoxytitanium compound and styrene. The radical polymerization of Ba-Ti complex alkoxide itself followed by hydrolysis was found to produce $BaTiO_3$ particles/polymer hybrids. Also, hybrid films including crystalline $BaTiO_3$ particles were synthesized from a mixture of Ba-Ti complex alkoxide and styrene. The dielectric properties of the hybrid films were investigated.

EXPERIMENTAL PROCEDURE

Barium metal and titanium tetraisopropoxide ($Ti(O^iC_3H_7)_4$) were commercially available. Methacryltriisopropoxytitanium (MTPT) was prepared from methacrylic anhydride ($(CH_2{=}C(CH_3)CO)_2O$) and titanium isopropoxide. Ethanol and 2-ethoxyethanol (EGMEE) were dried and distilled before use. Styrene was distilled before use.

MTPT and azobisisobutyronitrile (AIBN) were weighed in a glass tube, which was sealed under vacuum after repeating freezing-melting treatments in order to remove water and oxygen. The sealed glass capsule was heated at temperatures from 90 to 150°C for the period from 1 day to several days.

Figure 1 illustrates the representative processing scheme used for the synthesis of $BaTiO_3$/polymer hybrids. Barium metal was added in

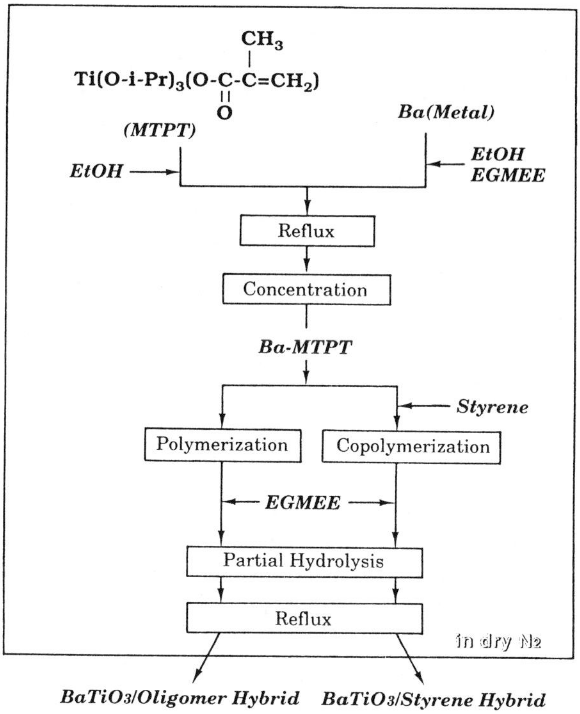

Fig. 1 Scheme for the processing of $BaTiO_3$/polymer hybrids

portion to a mixture of ethanol and 2-ethoxyethanol (EGMEE) (volume ratio of 20) at room temperature. An equimolar amount of MTPT was added and reacted at 80°C for 24h. The mixture (Ba-MTPT) was evaporated, and then polymerized at 100°C for 3 days using the same method as that of MTPT itself. Alternatively, styrene was added to the reaction mixture, which was copolymerized under the same conditions. The reaction product was dissolved in EtOH or EGMEE and hydrolyzed with water at 80°C for 24h.

The bulk $BaTiO_3$/polymer hybrid was placed between two steel plates with heating elements, and pressed fabricating films using an oil press. The film thickness ranged from 150 to 300 μm.

IR spectra of metallo-organic products were measured by the KBr method. Molecular weights were determined by GPC (Gel permeation chromatography) analysis. $BaTiO_3$/organic hybrid was analyzed by DTA-TG and X-ray diffraction analysis (XRD). The $BaTiO_3$ particles in an organic matrix were observed by TEM. The $BaTiO_3$/polymer hybrid was also analyzed by SAXS. The dielectric properties of the hybrid films were measured using LCR meter (HP 4275A). Au electrode was sputtered on the hybrid films prior to measurement. The D-E hysteresis measurement was conducted with a RT-66A ferroelectric test system (Radiant Technology).

RESULTS AND DISCUSSION

Polymerization of methacrylisopropoxytitanium

Methacrylisopropoxytitanium consists of metal-alkoxy bonds for condensation reaction and unsaturated double bonds for addition polymerization. The radical polymerization of methacryl-triisopropoxytitanium (MTPT) itself was confirmed by the change of IR spectra of polymerized products [6]. When azobisisobutyronitrile (AIBN) was used as a radical initiator, the polymerization proceeded with increasing the reaction time and temperature. The reaction product was a viscous liquid. At 150°C for 24h, the molecular weight was below 1000 [6].

Synthesis of $BaTiO_3$/oligomer hybrid from metallo-organic precursor

The formation of complex alkoxide from MTPT and barium alkoxide was confirmed by NMR spectroscopy [6]. The complex alkoxide (Ba-MTPT) was polymerized at 100°C for 3days, then hydrolyzed with various amounts of water at 80°C for 24h. The viscous product was analyzed by IR spectroscopy. The formation of carboxylate ($-COO^- M^+$) was confirmed by the presence of IR absorption at 1580 and 1400 cm^{-1}. The product also contained C-O absorptions at 1080 cm^{-1}. The change of ^1H NMR spectra also supports the formation of polymeric products [6]. The product was oligomeric, since the molecular weight was below 2000.

The polymeric hybrid formed from a Ba-MTPT/water ratio of 1.0 was analyzed by DTA-TG analysis. The TG curve of the hybrid showed a weight loss of 33% up to 700°C. The weight loss means that the hybrid includes 33 wt% organics. XRD analysis indicated that the hybrid was X-ray amorphous, and was crystallized to tetragonal $BaTiO_3$ on heating to 700°C.

Figure 2 reveals the microstructure of the $BaTiO_3$/oligomer. The black particles were dispersed in a matrix as shown in Fig. 2a. The electron diffraction of black particles shown in Fig. 2b reveals the presence of crystalline phase. The diffraction pattern was in good agreement with that of $BaTiO_3$. The particle diameter was below 8 nm. The cumulative frequency of particles showed that a median diameter was 2 nm. Also, the diameters of $BaTiO_3$ in the polymer were found to comprise from 2.7 nm (82 wt%) and 8.1 nm (15%) by SAXS.

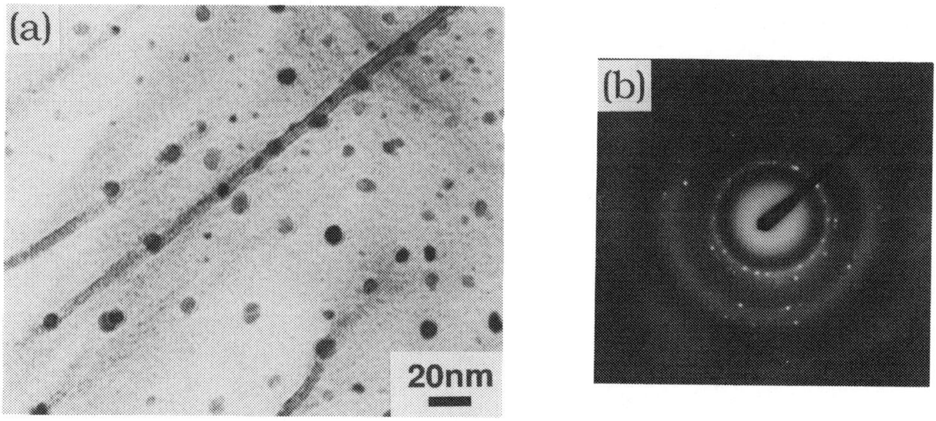

Fig. 2 Microstructure of $BaTiO_3$/polymer hybrids with electron diffraction of black particles.

Synthesis of $BaTiO_3$/copolymer hybrid with styrene

Styrene was added to the Ba-Ti complex alkoxide in order to increase the formability of the hybrid polymer. The mixture of styrene and MTPT was polymerized using AIBN under the same conditions as those of Ba-Ti alkoxide. The molar ratios of styrene and Ba-MTPT were 0.5, 1.0 and 2.0. The molecular weights of polymerized products analyzed by GPC were 975 (molar ratio; 0.5), 1030 (1.0) and 3490 (2.0). The molecular weight of the products increased with increasing ratio of styrene to Ba-MTPT. The products were soluble in 2-ethoxyethanol, and used for the subsequent hydrolysis for the formation of $BaTiO_3$ particles in the polymer matrix.

After the mixture solution was hydrolyzed and refluxed under the same conditions, the products were analyzed by TEM. The particles dispersed in the polymer matrix were less than 10 nm in diameter, and were found to be crystalline $BaTiO_3$ particles by SAD.

The IR spectrum of the $BaTiO_3$/oligomer hybrid and the $BaTiO_3$/styrene hybrid are shown in Fig. 3. The spectrum of the latter (Fig. 3b) consists of the characteristic absorptions of the $BaTiO_3$/oligomer hybrid at 1580 and 1400 cm^{-1} (Fig. 3a) and those of polystyrene (1800, 1870 and 1940 cm^{-1}).

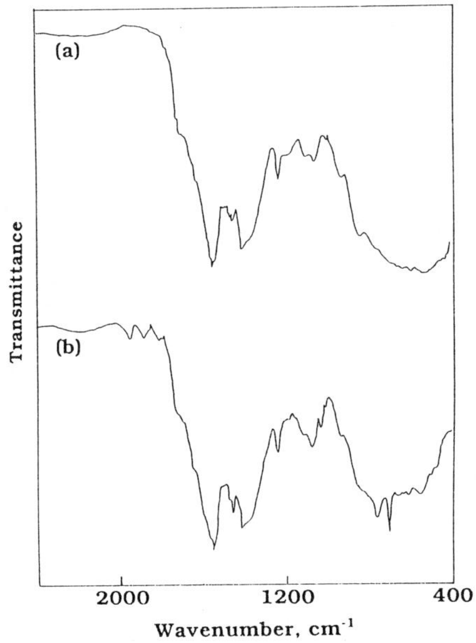

Fig. 3 IR spectra of (a) $BaTiO_3$/oligomer, and (b) $BaTiO_3$/styrene hybrid.

The $BaTiO_3$/polymer hybrid obtained by copolymerization between $BaTiO_3$/oligomer and styrene is composed of $BaTiO_3$ particles and polymeric backbones. The polymeric part has a possible structure unit shown in eq. (1), and includes unprecipitated $BaTiO_3$ particles of quite small sizes.

$$CH_2{=}C(COOM(OR)_n){-}CH_3 \xrightarrow{\text{styrene}} -(CH_2{-}C(COOM(OR)_n){-}CH_3)_n{-}\underbrace{(CH_2{-}CH{-}C_6H_5)_m}_{\text{styrene part}}{-} \quad (1)$$

R=H, alkyl, M'

Dielectric properties of $BaTiO_3$/styrene hybrid

Figure 4 shows the dependence of dielectric constant (ε) of the hybrid (Ba-MTPT/styrene=1.0) on the amount of water used for the hydrolysis. The ε increases with increasing amount of added water for hydrolysis. The ε reached constant at water amounts over 0.5 equiv. to Ba-MTPT. The increase of dielectric constants is considered to result from the increase of size and crystalline regularity of $BaTiO_3$ particles precipitated out from Ba-Ti alkoxide in the polymer matrix.

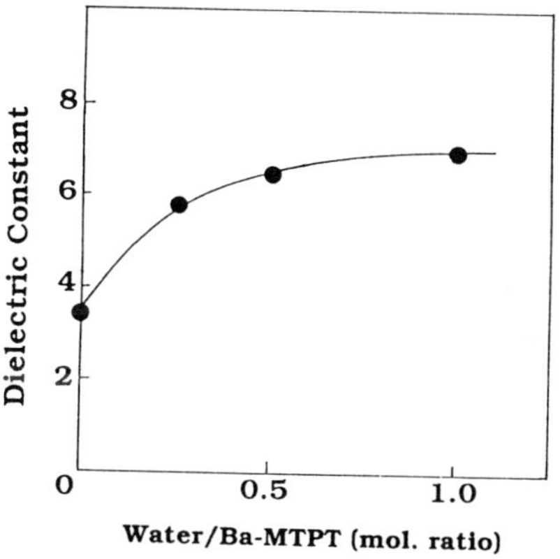

Fig. 4 Change in dielectric constant of $BaTiO_3$/styrene hybrid with amount of added water at 100kHz.

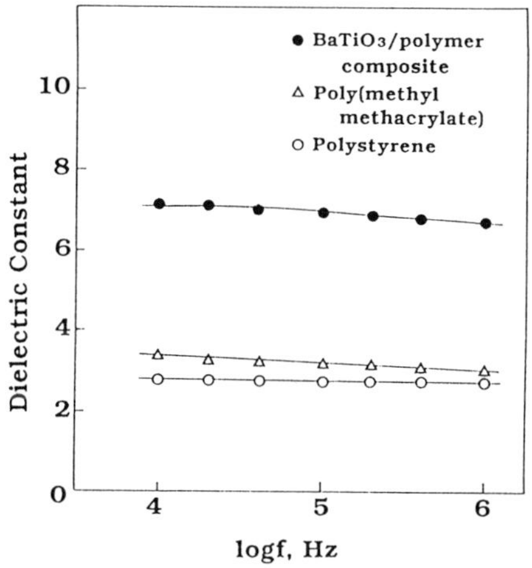

Fig. 5 Dependence of dielectric constant of $BaTiO_3$/ styrene hybrid on frequency at room temperature.

The relation between dielectric constant of $BaTiO_3$/styrene hybrid film (Ba-MTPT/water=1.0) and frequency is shown in Fig. 5. Figure 5 also shows the dielectric constants of polystyrene and polymethylmethacrylate films. The ε is almost constant of about 7, and decreased slightly at higher frequency. The methyl methacrylate polymer film shows a similar decrease, whlie ε of polystyrene dose not show any decrease. The reason for the decrease is, therefore, attributed to the presence of polar carboxylate group in the hybrid film. The loss tangent was from 0.025 to 0.03 over the measured frequency.

Figure 6 shows the changes in dielectric constant of $BaTiO_3$/styrene hybrid and polystyrene with temperature at 100kHz. The hybrid film was synthesized from a molar ratio of styrene to Ba-MTPT of 1.0 at 1.0 equiv. water. The dielectric constant is 7, and shows no maximum around 120°C. The critical size of $BaTiO_3$ particles for tetragonality is reported to range from 80 to 120 nm [7,8]. The present hybrids includes much smaller size of $BaTiO_3$ particle. Therefore, the quite small size of $BaTiO_3$ particles is responsible for the absence of dielectric maximum around 120°C, relating to the Curie temperature.

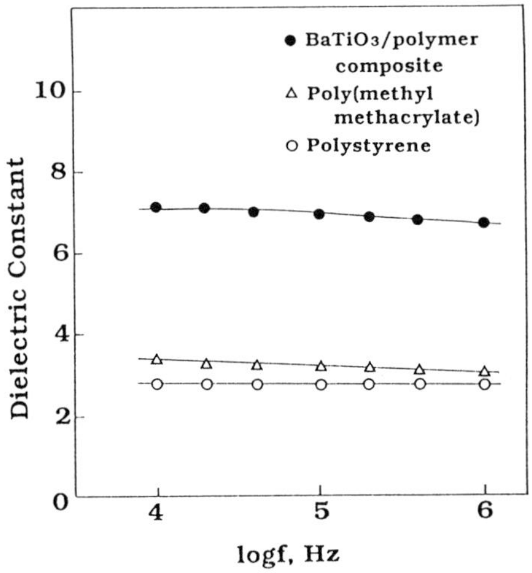

Fig. 6 Changes in dielectric constant of $BaTiO_3$/styrene hybrid and polystyrene with temperature measured at 100kHz.

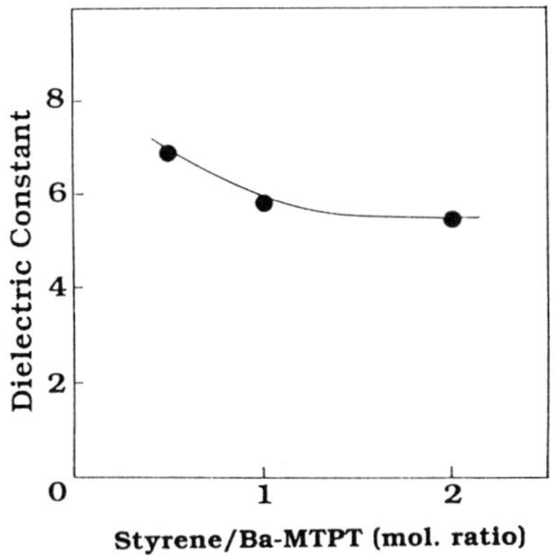

Fig. 7 Change in dielectric constant of $BaTiO_3$/ styrene hybrid with amount of styrene.

Figure 7 shows the change in dielectric constant of the hybrid with styrene (Ba-MTPT/water=1.0) at 100kHz. The ε decreases with increasing amount of styrene from 0.5 to 1.0. The dielectric constant of $BaTiO_3$ particles in the hybrid was estimated using the relation of dielectric constant for two phase materials [9]. The dielectric constant of $BaTiO_3$ was found to be about 20, when the ε of polystyrene is 2.28. The quite small value of the present $BaTiO_3$ compared with bulk $BaTiO_3$ is mainly due to the small particle size.

Figure 8 illustrates D-E hysteresis curves of the hybrid films with the different styrene/Ba-MTPT ratios (water amount, 1.0 equiv). The hysteresis measurements were conducted for the films under the same measurement conditions. The hybrid films reveal hysteresis loops, while a polystyrene film showed a linear relation between D and E. The remanent polarization increases with an increase of Ba-MTPT in the hybrid, and is related to the amount of $BaTiO_3$ in the hybrid film. The hysteresis is considered to result from the interaction of crystalline superparaelectric $BaTiO_3$ particles in the polymer matrix.

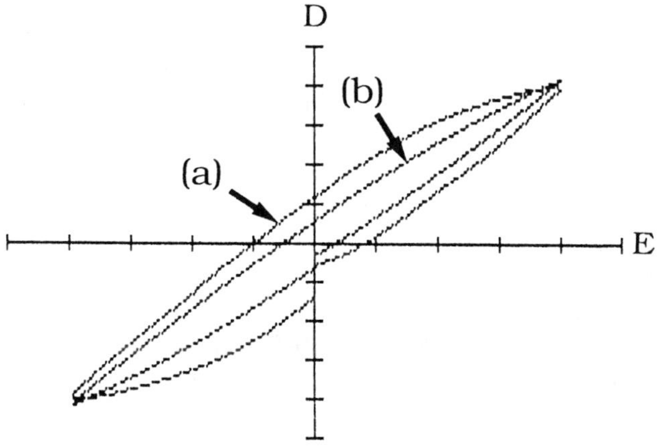

Fig. 8 D-E hysteresis of $BaTiO_3$/styrene hybrid.
(a) styrene/Ba-MTPT=0.5, (b) styrene/Ba-MTPT=1.0.
(X div.=5.0 V, Y div.=2.0μC/m^2)

CONCLUSIONS

$BaTiO_3$/polymer nano-composite was successfully synthesized from methacryltriisopropoxytitanium (MTPT) and barium alkoxide using organic polymerization combined with hydrolysis. The metallo-organic precursor with a Ba/Ti ratio of 1 yielded $BaTiO_3$ particles by in-situ crystallization at 80°C. Crystalline $BaTiO_3$ particles less than 3 nm were successfully dispersed in a polymer matrix. The dielectric constant of $BaTiO_3$/styrene hybrid was dependent on the amount of hydrolysis water and on the ratio of styrene/Ba-MTPT. The D-E hysteresis was observed for the $BaTiO_3$/styrene hybrid film.

References

1. H.Schmidt, "Organically Modified Silicates by the Sol-Gel Process", *Mater. Res. Soc. Symp. Proc.* **32** edited by J.C.Brinker, D.E.Clark and D.R.Uhlrich (Elsevier, Amsterdam, 1984) p.327-35.
2. H.Schmidt, "New Type of Non-crystalline Solids between Inorganic and Organic Materials", *J. Non-Cryst. Solid*, **73** [1-3] 681-91 (1985).
3. G.L.Wilkes, B.Orler and H-H. Huang, "Ceramers: Hybrid Materials Incorporating Polymer/Oligomeric Species into Inorganic Glasses Utilizing a Sol-Gel Approach", *Polym. Prep.* **26** [2] 300-01 (1985).
4. H.Schmidt, "Processing and Optical Properties of Inorganic-Organic Composites", *Chemical Processing of Advanced Materials* edited by L.L.Hench and J.K.West, (John Wiley & Sons, New York, 1992) p.727-35.
5. C.Sanchez and M.Inn, "Molecular Design of Alkoxide Precursors for the Synthesis of Hybrid Organic-Inorganic Gels", *J. Non-Cryst. Solid* **147&148**, 1-12 (1992).
6. T.Yogo, K.Kikuta, S.Yamada and S.Hirano, "Synthesis of Barium Titanate/Polymer Composite from Metal Alkoxide", *J. Sol-Gel Sci. Tech.* **2** [1-3] 175-79 (1994).
7. J.R.Slack and J.C.Burfoot, "Electrical Properties of Flash Evaporated Ferroelectric $BaTiO_3$ Thin Films", *J. Phys. C: Solid St. Phys.,* **4** [8] 898-909 (1971).
8. K.Uchino, E.Sadanaga and T.Hirose, "Dependence of the Crystal Structure on Particle Size in Barium Titanate", *J. Amer. Ceram. Soc.,* **72** [8] 1555-58 (1989).
9. T.Furukawa, K.Fujino and E.Fukuda, *Jpn. J. Appl. Phys.* **15** [11] 2119-29 (1976).

SOL-GEL PROCESSING OF $Pb(Zr_{0.52},Ti_{0.48})O_3$ THIN FILMS : DOPANT COMPENSATION MECHANISM AND LEAKAGE CURRENT CHARACTERISTICS

C.K. BARLINGAY and S.K. DEY
Department of Chemical, Bio and Materials Engineering and
Centers for Solid State Science/Electronics Research,
Arizona State University, Tempe, AZ 85287-6006.

Abstract:

Dopant compensation mechanism of sol-gel derived $Pb(Zr_{0.52},Ti_{0.48})O_3$ or PZT thin films with 2-5 mole percent niobium (Nb) dopant was studied. Transmission electron microscopy (TEM) showed that the Nb-doped composition batched according to the Ti vacancy-model, was single phase. This indicated that niobium donors were compensated by titanium vacancies at the B-site. The beneficial effect of donors in single phase composition and detrimental effects of grain boundary second phases and acceptors on the leakage currents are reported. The single phase Nb-doped (5%) PZT thin film exhibited a low leakage current of 1×10^{-7} A/cm^2 at 1.2 MV/cm and a high breakdown strength of 3 MV/cm. Additionally, the logarithmic plot of steady state conductivity versus the reciprocal of temperature (T), exhibited a change in the activation energy at 155 °C. The activation energies of 1.2 eV at T >155 °C and 0.1 - 0.2 eV at T < 155 °C for the steady state conductivity were observed in Nb doped PZT thin films.

1. Introduction

It is well known that both microstructure and solid state chemistry influence the electrical properties of perovskite materials.(1-5)

To the extent authorized under the laws of the United States of America, all copyright interests in this publication are the property of The American Ceramic Society. Any duplication, reproduction, or republication of this publication or any part thereof, without the express written consent of The American Ceramic Society or fee paid to the Copyright Clearance Center, is prohibited.

Surprisingly, to date, distinct effects of aliovalent acceptor or donor dopants have not been reported(6,7) in the Pb-based perovskite thin-films. This probably is due to the lack of stoichiometry and microstructure control.

For a bulk perovskite (ABO_3) material, it is well documented(1,8) that the Schottky disorder involves the formation of three types of ionized vacancies, V_A'', V_B'''', and $V_O^{\bullet\bullet}$. At equilibrium, the concentration of these defects are related through the mass-action constant equation 1.

$$[V_A''][V_B''''][V_O^{\bullet\bullet}]^3 = \text{constant} \qquad (1)$$

where the brackets denote the concentration of enclosed species. The super-oxidation of $Pb(Zr,Ti)O_3$ or PZT, as shown in equation 2, results in p-type conductivity.

$$V_O^{\bullet\bullet} + 1/2\, O_2 \dashrightarrow O_O + 2\dot{h} \qquad (2)$$

The site occupancies of stoichiometric (A/B=1) PZT ceramics at high P_{O_2} and high donor (e.g., Nb) concentrations are given by(1):

$$PbNb_x\,(Zr,Ti)_{1-x}\,O_{3+x/2} \dashrightarrow$$

$$[Pb_{1-x/2}\;V_{Pb}''{}_{x/2}]_A[Nb^{\bullet}_x\,(Zr,Ti)_{1-x}]_BO_3 + x/2\,PbO \qquad (3)$$

if compensation occurs by the generation of A-site vacancies (i.e., V_{Pb}''). Thus, the concentration of oxygen vacancies (due to incidental acceptor impurities) and electronic charge carriers (due to super-oxidation) are suppressed based on Schottky equilibrium. These factors render the bulk material insulating.

Recent SIMS, AES, SEM and TEM studies(9) have indicated that when PZT thin films are deposited on the $Pt/Ti/SiO_2/Si$ structures, severe diffusion of Ti (glue layer) through the Pt and Pb diffusion into the Pt/Ti layers occur during heat-treatment at temperatures between 300-700°C. Therefore, in the event of PbO loss either to the air or substrate would require a high donor concentration (within the solubility limit) in order to obtain phase pure insulating films. In essence, donor doping in phase pure perovskite PZT thin films is not successful if loss of PbO is greater than the donor concentration. Indeed, Raymond et al.[10] have reported that a higher amount of donor is required to suppress the $V_O^{\cdot\cdot}$ concentration in the presence of lead loss. When loss of PbO is so significant as to form several phases, studies to determine the effects of donor doping are meaningless.

This paper reports on the: 1) details of crack-free sol-gel thin-film processing, 2) reasons for addition of formamide as drying control chemical additive (DCCA), 3) defect compensation mechanism in donor-doped (2-5 mole %) $Pb(Zr_{0.52},Ti_{0.48})O_3$ thin films with deliberately added 10 mole % excess PbO as determined by transmission electron microscopy (TEM), and 4) effects of donors and acceptors on the microstructure and the leakage current characteristics of PZT(52/48) thin-film system.

2. Experimental Details

2.1 Thin-film fabrication

The standard Schlenk technique was used to synthesize the Pb-Nb-Zr--Ti or PNZT precursor complex. This is a modification of process originally reported by Gurkovich and Blum.[11] A flow diagram for precursor synthesis is illustrated in Figure 1. Initially, zirconium n-propoxide, titanium isopropoxide and niobium ethoxide were individually converted to their respective methoxyethoxides by atmospheric distillation (Equations 4a-c).

$$Ti^{i}(OC_3H_7)_4 + CH_3OCH_2CH_2OH \longrightarrow Ti(OCH_2CH_2OCH_3)_4 + C_3H_7OH \quad (4a)$$

$$Zr^{n}(OC_3H_7)_4 + CH_3OCH_2CH_2OH \longrightarrow Zr(OCH_2CH_2OCH_3)_4 + C_3H_7OH \quad (4b)$$

$$Nb\,(OC_3H_7)_5 + CH_3OCH_2CH_2OH \longrightarrow$$

$$Nb\,(OCH_2CH_2OCH_3)_5 + C_3H_7OH \qquad (4c)$$

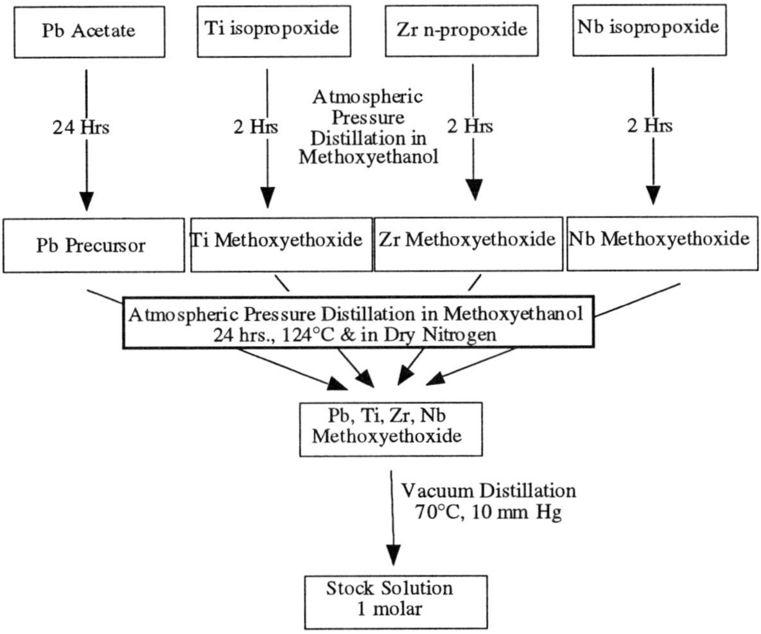

Figure 1. A flow diagram for Pb $(Nb,Zr,Ti)O_3$ precursor synthesis.

Next, Zr, Sn, and Ti methoxyethoxides were reacted with anhydrous Pb-acetate and appropriate amounts of niobium methoxyethoxide were introduced under dry nitrogen. The 10% excess PbO necessary to compensate for the aforementioned loss during processing was also added at this stage. Details of stock solution preparation were reported previously.[9,12,13]

The steps involved in thin-film fabrication are indicated in the Figure 2. A 50 cc hydrolysis solution contained 0.2 moles of deionized water and 0.03 moles of NH_4OH in methoxyethanol solvent. The precursor solution was then hydrolyzed by mixing a 1 molar stock solution with hydrolysis solution in the ratio of 2:1. The hydrolyzed solution, contained 2.0 moles of H_2O per mole of PZT alkoxide and 0.2 molar basic catalyst (NH_4OH) which was then aged for 24 hours. Next, methoxyethanol and drying control chemical additive (formamide) were mixed before obtaining bulk gels or thin films.

The reasons for addition of formamide as the drying control chemical additive (DCCA) are described hereinafter. The physics and chemistry of thin film formation is characterized by a concentration induced aggregation process. Here gelation and drying processes occur simultaneously. However, extensive shearing and too high an evaporation rate (and hence, drying rate) with respect to condensation rate (that promotes gelation), traps solvent and causes defects that give rise to pin-holes and streaks. The use of formamide in our system is to decrease the evaporation or rate (i.e., to increase the drying time) with respect to gelation rate during the spin-casting process (which is about 60-120 seconds). This results in defect-free films that have high electrical breakdown strengths when heat treated. The overall volumetric proportion of stock solution, hydrolysis solution, methoxyethanol, and formamide in this spin-casting solution was 2:1:0.5:0.5, respectively. The thin films were obtained by spin casting onto (as-received) three inch diameter Pt/Ti/SiO_2/Si wafers at 2000 rpm for 80 seconds. A typical thickness of 1000 Å per layer was obtained.

Each layer is heat treated at 400°C for 1 second and 350°C for 60 seconds per layer followed by viscous sintering at 350°C for 15 minutes after the final layer. The films subjected to such treatments are referred to as amorphous films. The final crystallization and densification of the PZT thin films were carried out at 700 °C for 5 minutes (referred to as crystalline films) using an AG Associates Heat Pulse 610 rapid thermal annealing (RTA) furnace.

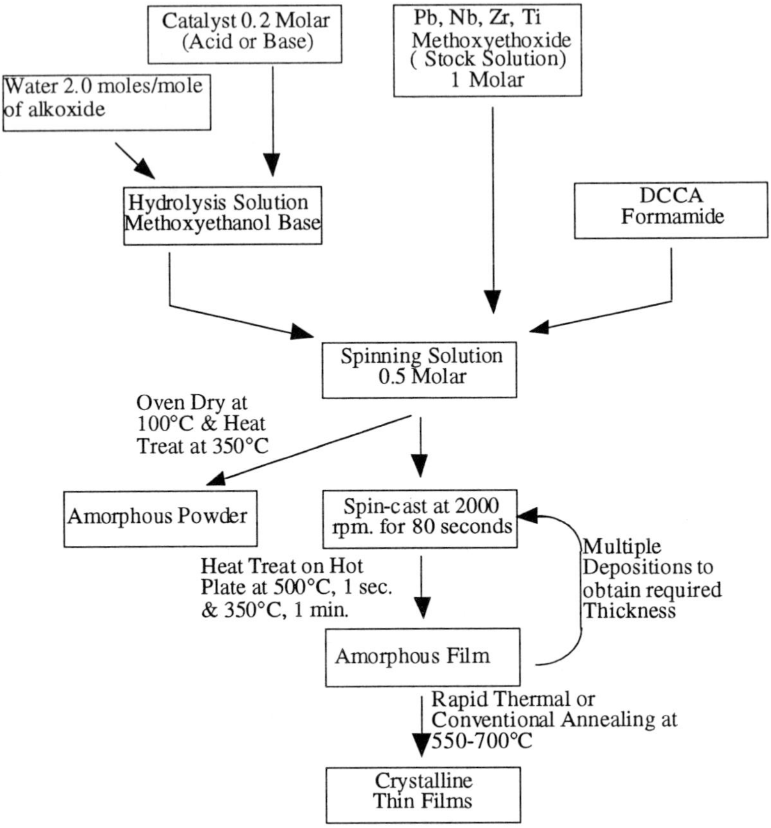

Figure 2. A flow diagram for the fabrication of $Pb(Nb,Zr,Ti)O_3$ thin films.

2.2 Characterization

For carrying out Transmission Electron Microscopy (TEM), specimens were prepared by ultrasonically cutting 3 mm disks from $PZT/Pt/Ti/SiO_2/Si$ wafers. The disks were then polished and dimpled from the back side, followed by Ar^+ ion milling (4-5 kV,

1mA) at an incident angle of 5-15 degrees. Circular mica shields were used to protect the film side from contamination during ion milling.

Microstructural examinations on PZT thin films including electron diffraction, energy dispersive X-ray analysis (EDX), transmission electron microscopy (TEM) were carried out on a JOEL 2000 FX side entry microscopes equipped with an energy dispersive x-ray spectrometer (KEVEX) (for EDX).

The capacitor area was defined by top Pt electrodes, deposited by using TEM copper grids as shadow masks in an e-beam evaporator. The polarization-field (P-E) properties were evaluated using the standard Sawyer-Tower circuit in conjunction with a HP54503A Digitizing Oscilloscope and Wavetek 191 20 MHz Pulse/Function Generator. The leakage current characteristics (I-V) were measured up to an applied field of 1.2 MV/cm on a femto-ampere current meter (Tektronix 372). The thin-film capacitors were poled with a 400 kV/cm dc field for 10 minutes prior to the I-V measurements. This eliminated any switching current contribution or transient effects to the leakage current.(5) The reported data correspond to a 5 second waiting period after the application of each voltage step. The temperature dependence of the steady state leakage current was measured on a temperature controlled heated chuck. The breakdown strengths were determined during continuous voltage stress with Vstep and delay of 0.25 V and 5 s, respectively.

3. Results and Discussions

The compositions with aliovalent dopants (Table I) were batched assuming A, B, and O site vacancy models as follows:

1) A-site vacancy model: $Pb_{1-x/2}\,[Zr_{0.52}Ti_{0.48}]_{1-x}\left[Nb^{\bullet}_{Ti}\right]_x\,O_3$, where X = 2, & 5 mole %.

2) B-Site vacancy model: $Pb[Zr_{0.52}Ti_{0.48}]_{1-5x/4}\left[Nb^{\bullet}_{Ti}\right]_x\,O_3$, where X = 5%

3) Oxygen vacancy model: $Pb[Zr_{0.52}Ti_{0.48}]_{1-x}[Fe'_{Ti}]_x O_{3-x/2}$, where X= 2%

All compositions were batched such that they would be single phase, if the assumed defect model was operating. The A to B-site ratio, and TEM results are summarized in Table I, along with the designation of the compositions. For example, the A-site vacancy compensated composition (the first formula on the right hand side of equation 2 with the vacancy discounted) was selected as the one that would be a single phase if charge compensation was due to V''_{Pb}. Therefore, the batched Pb/[Nb + Zr + Ti] ratio of A2 composition is 0.99; i.e., non-stoichiometric and lead deficient.

Table I.

Vacancy compensation model, sample label, batched composition, along with A/B ratio and observed phases in the TEM for acceptor and donor doped PZT thin films.

Model	Sample Label	Batched Composition	A/B Ratio	TEM Results
A-site	A2	$Pb_{0.99}[Zr_{0.52}Ti_{0.48}]_{0.98}[Nb^{\bullet}_{Ti}]_{0.02}O_3$	0.99	Lead deficient, GB phase
A-site	A5	$Pb_{0.975}[Zr_{0.52}Ti_{0.48}]_{0.95}[Nb^{\bullet}_{Ti}]_{0.05}O_3$	0.975	Lead deficient, GB phase
B-site	B5	$Pb[Zr_{0.52}Ti_{0.48}]_{0.9375}[Nb^{\bullet}_{Ti}]_{0.05}O_3$	1.013	no GB phase
Oxygen	O2	$Pb[Zr_{0.52}Ti_{0.48}]_{0.98}[Fe'_{Ti}]_{0.02}O_3$	1	no GB phase, embedded porosity

Figures 3 (a) and (b) illustrate planar microstructures of Nb donor-doped A2 and A5 films, respectively. The microstructures of these films consist of perovskite grains surrounded by a distinct grain boundary (GB) second phase. The observed second phase at the GBs was determined by EDX to be Zr and Ti rich. Additionally, the electron diffraction pattern indicated that this nano-crystalline GB phase was a lead-deficient cubic pyrochlore ($Pb_{2-\delta}(Zr,Ti)_2O_{7-x}$) with a lattice parameter of 10.4 Å.[9,14] The GB phase in A2 was observed predominantly at the triple points, whereas in A5 the GB phase surrounding the perovskite grains was continuous. For the higher Nb doped film (A5), the amount of second phase was higher (compare Figures 3 a & b) but the grain size (0.8-1.5 μm) remained the same. Interestingly, B5 (donor-doped composition batched as non-stoichiometric and Ti deficient) thin film exhibited a single phase (perovskite) microstructure (Figure 3 c) with an average grain size of 0.8-1.2 μm. The above result indicate that donors in sol-gel derived PZT thin films are compensated by $V_{Ti}^{''''}$ rather than $V_{Pb}^{''}$ since a single phase microstructure is exhibited only by B5. Thus, the high B/A ratio in the GB phase of A2 and A5 can be explained by the exsolved B-site cations if compensation occurs according to the B-site vacancy-model. Additionally, the amount of second phase is higher in A5 than A2 as expected since the Nb content in the former composition is higher. Figure 3d illustrates phase pure perovskite microstructure of O2 thin film (acceptor-doped), where embedded porosity within the grains with clean grain boundaries was observed.

The results of this study are in concurrence with an earlier report [3] of $V_{Ti}^{''''}$ as the compensating defect in Nb-doped (> 5 mole %) $BaTiO_3$. Contrary to the above observations, $V_{Pb}^{''}$ has been reported to be the compensating defect in Nb-doped, bulk PZT ceramics.[1,2] This discrepancy may be explained by the increased activity of Pb (since excess PbO was added to compensate for the loss during processing) and therefore, the observed $V_{Ti}^{''''}$ compensating defect model. Indeed, Atkin and Fulrath[15] have shown that at a constant oxygen partial pressure, the concentration of $V_{Pb}^{''}$ decreases and $V_{Ti}^{''''}$ increases with an increase in the activity of PbO. The observed $V_{Ti}^{''''}$ model may also be inherent for highly donor-doped PZT, but simulation studies, as carried out by Catlow et al. for the $BaTiO_3$ system,[16] must be done.

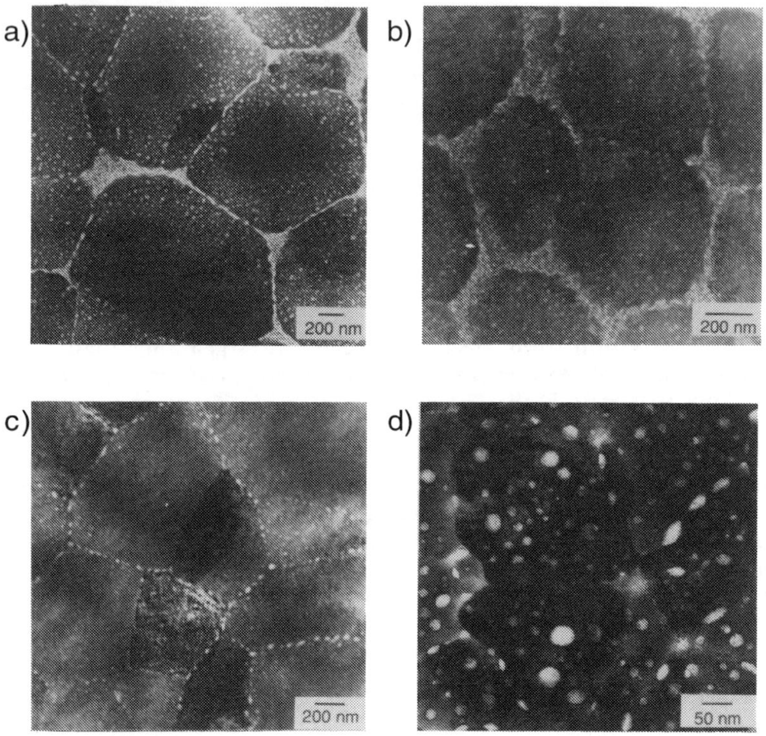

Figure 3: TEM planar bright field images of PNZT thin films batched assuming vacancy models a) A-site 2% Nb, b) A-site 5% Nb, c) B-site 5% Nb, and d) Oxygen 2% Fe.

The electrical properties of acceptor and donor doped PZT thin films are reported below. Stable and fully saturated hysteresis loops (uncompensated) were obtained at 100 Hz and 300 kV/cm applied field (Figure 4). The hysteresis properties of O2 (2% Fe-doped PZT) were similar to undoped PZT. The remnant polarization, P_r, maximum polarization, P_{max} , and coercive field, E_c were in the ranges of 24-25 $\mu C/cm^2$, 44-45 $\mu C/cm^2$, and 44-50 kV/cm, respectively, The single phase B5 (5% Nb-doped) PZT exhibited P_r, P_{max}, and E_c in the ranges of 15-17 $\mu C/cm^2$, 32-34 $\mu C/cm^2$, and 40-45 kV/cm, respectively. In A2 and A5, the P_r, and P_{max}, were 30-35 percent lower than B5 due to dilution effects.

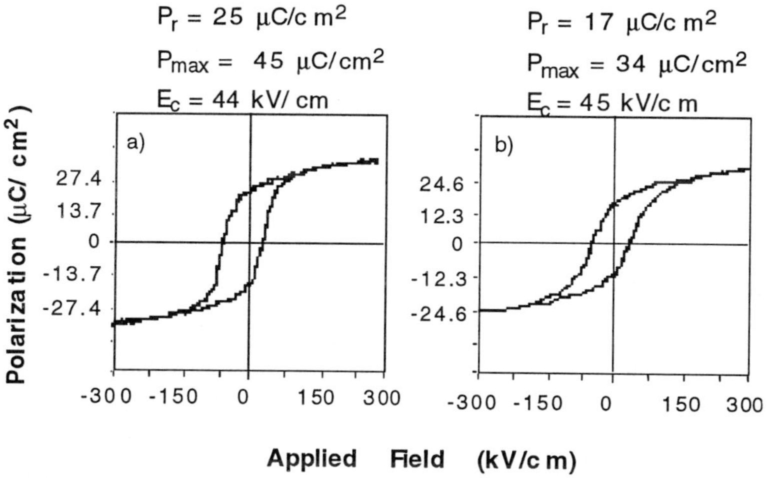

Figure 4: Hysteresis loops of PZT thin films a) O2 i.e., 2% acceptor (Fe) doped, and b) B5 i.e., 5% donor (Nb) B-site compensated.

The effects of microstructure and solid state chemistry on the leakage currents of PZT(52/48) thin films are reflected in Figure 5. Let's consider the phase pure perovskite films (i.e., O2 and B5) first, where the solid state chemistry plays an important role. The O2 (2 % Fe-doped) PZT films exhibited higher leakage current density ($J_L \approx 10^{-7}$ A/cm^2) at low applied fields (< 200 kV/cm) when compared to the donor doped B5 composition. Additionally, the breakdown strength (400 kV/cm) of O2 is approximately an order of magnitude lower than breakdown strength of B5 (≈ 3 MV/cm). The observed results for O2 are consistent with the $V_O^{\cdot\cdot}$ model. Thus, a high initial $V_O^{\cdot\cdot}$ concentration leads to a field-induced concentration polarization of $V_O^{\cdot\cdot}$ with a pronounced increase in the electrical conductivity.(5) If the initial concentration of $V_O^{\cdot\cdot}$ is low, the leakage current should be low. Indeed, the low J_L (10^{-7} A/cm^2 at 1.2 MV/cm) and a high breakdown strength (≥3 MV/cm) for B5 are as expected since the $V_O^{\cdot\cdot}$ concentration is suppressed in this composition.

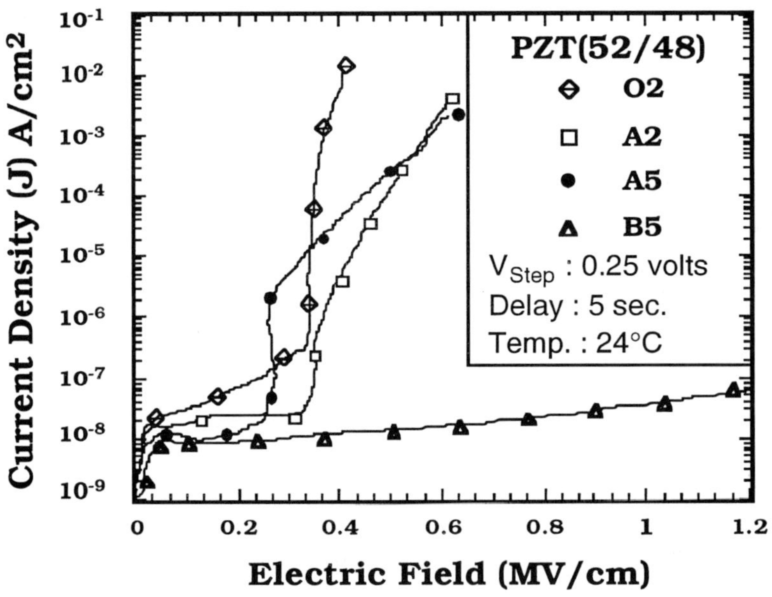

Figure 5: I-V characteristics of acceptor and donor doped PZT thin films (0.46 μm) measured on 90x90 μm^2 capacitor areas.

The leakage current in diphasic A2 & A5 thin films with GB phases are influenced, additionally, by extrinsic (microstructural) effects. For example, A5 which has a higher amount of GB second phase compared to A2, showed low J_L ($\approx 10^{-8}$ A/cm^2) below 200 kV/cm but a much higher leakage above 200 kV/cm. However, at high fields (> 500 kV/cm), J_L of both (A2 & A5) approached similar limiting value (1-3 x 10^{-3} A/cm^2). These results can be explained by considering the microstructure and conductivities (grain and GB) of A2 and A5 films. Since A2 and A5 exhibit columnar grain structures (not shown here), parallel leakage paths through the grain, and GB second phases are present. According to parallel conductivity mixing rule:

$$\sigma_{Total} = v_g \sigma_g + v_{GB}\, \sigma_{GB} \qquad (5)$$

where, σ_g and σ_{GB} are conductivities of insulating grain and GB second phases with volume fractions of v_g and v_{GB}, respectively.

At low fields (< 200 kV/cm), the grain conductivity ($v_g\sigma_g$ term) dominates J_L because the volume fraction of GB second phase is small. Since the grains of A5 film, with a higher donor concentration, is more insulating than A2 [i.e., $(\sigma_g)_{A5} < (\sigma_g)_{A2}$], the J_L of A5 ($\approx 10^{-8}$ A/cm^2) at 200 kV/cm is lower than A2 (3 x 10^{-8} A/cm^2). Additionally, similar J_L values of A5 and B5 at fields less than 200 kV/cm can be explained by the role of Nb-donors, in equivalent amounts within the grain, coupled with the minor influence of the GB phase on the conductivity in A5.

At high fields (> 300 kV/cm), however, the conductivity of the GB second phase begins to contribute to the total J_L. The conductivity enhancement at 260 kV/cm for A5, having a higher v_{GB}, is expected. A similar conductivity enhancement in A2 film occurs at a higher field (340 kV/cm) because v_{GB} in A2 is lower compared to A5. Note that σ_{GB} contributes to J_L, especially at high fields, since the GB phases are non-stoichiometric. At higher fields (> 300 kV/cm), the conductivity of A2 and A5 compositions is a complex function of $v_g\sigma_g$ and $v_{GB}\sigma_{GB}$ leading to a similar limiting value.

To gain further insight into the leakage current mechanism, the steady state leakage current at different temperatures was measured with 100 kV/cm applied field. A typical time-dependent dc current diminished by two orders of magnitude in the first 60 second and eventually reached a steady state value. Similar trends have been observed in both ceramic and thin film $BaTiO_3$ and $Pb(Zr,Ti)O_3$.[5,17] Numerous mechanisms[18] have been proposed to explain this behavior. Most of them such as electrode polarization, and space charge polarization, invoke the ionic motion of charge. In this study steady state current shifted to higher values with increasing temperature. To determine the activation energy for the steady state part of the leakage current logarithmic plot as shown in Figure 6 was made.

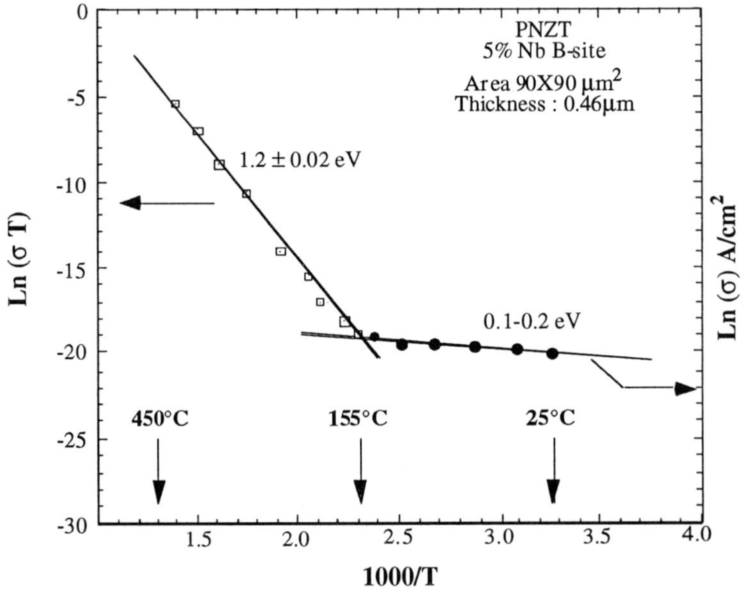

Figure 6. Logarithmic plot of steady state conductivity vs. I/T for B5 composition.

The plot (Figure 6) exhibits a change in slope at 155 °C which is indicative of a change in activation energy for the steady state conductivity. The activation energies of 1.2 eV at T >155 °C and 0.1 - 0.2 eV at T < 155 °C were observed. The activation energies for PZT as reported previously by Dih-Fulrath[19], Gerson and Jaffe[20] and Takahashi[21] are 1.41 eV, 0.89, 1.054 eV, respectively. Additionally, for $Pb_{0.97}[(Zr_{0.52},Ti_{0.48})_{0.94}Nb_{0.06}]O_3$ ceramic, a similar change in activation energy at 180 °C was observed by Ikeda et al.[22] The presence of acceptors are known to decrease the activation energy for conductivity, while the donors increase it. The increased leakage current and decreased breakdown strength for O2 composition indicate that oxygen vacancies play a dominant role in the leakage current mechanism at high temperature and high fields. However, further studies necessary to identify the leakage current mechanism in these PZT thin films.

4. Conclusions

An increased amount of a lead-deficient second phase was observed for thin film compositions batched according to the A-site vacancy model when the amount of Nb was increased. A quick inspection of the leakage characteristics of Nb-doped (A2 and A5) thin films do not show an improvement over the acceptor doped (O2) films; observation similar to that reported in the literature. This is a direct result of the diphasic microstructures exhibited by A2 and A5 i.e., the conducting GB phases enhance the leakage current at high fields. By compensating for the lead loss to the substrate, the beneficial effect of donors and the detrimental effect of acceptors can be observed even in thin films. The B5 composition with an A/B ratio of 1.013, was a phase pure perovskite film. This indicated that donors were compensated by $V_{Ti}^{''''}$ at the B-site. This mode of compensation was probably due to the increased activity of lead. As a result, a J_L of 10^{-7} A/cm^2 at 1.2 MV/cm and the breakdown strength of ≥ 3 MV/cm were observed. The effectiveness of Nb donor to reduce the leakage current was evident. For acceptor-doped O2 film with no GB phase, J_L was 10^{-7} A/cm^2 at 200 kV/cm but increased to 10^{-3} A/cm^2 at 400 kV/cm. Additionally, the logarithmic plot of steady state conductivity versus the reciprocal of temperature (T) for these single phase compositions, exhibited a change in slope at 155 °C. The activation energies of 1.2 eV at T >155 °C and 0.2 eV at T < 155 °C were observed in Nb doped PZT thin films.

References:

1. B. Jaffe, W. R. Cook and H. Jaffe, Piezoelectric Ceramics. Academic, New York, 1970.

2. T. Ikeda, Y. Tanaka, T. Ayakawa, H. Naoke, "Precipitation of Zirconia Phase in Niobium-Modified Ceramics of Lead Zirconate-Titanate, "Jap. J. Appl. Phys., vol. 3 [10], pp. 581-87 (1964).

3. H. M. Chan, M. P. Harmer, D. M. Smyth, "Compensating defects in highly donor-doped $BaTiO_3$", J. Am. Ceram. Soc., vol. 69 [6], pp. 507-10 (1986).

4. D. Henning and K. H. Härdtl, "Distribution of A-site and B-site Vacancies in (Pb,La) $(Ti,Zr)O_3$ ceramics," Phys. Stat. Sol. (a) vol. 3, pp. 465-76 (1970).

5. R. Waser, T. Baiatu, and K. H. Härdtl, "dc Electrical Degradation of Perovskite-type Titanates: I, Ceramics, II, Single Crystal, and III: Model," J. Am. Ceram. Soc. vol. 73 [6], pp. 1645-73 (1990).

6. Daniel F. Ryder and Narayan K. Raman, "Sol-Gel Processing of Nb-Doped $Pb(Zr, Ti)O_3$ Thin Films for Ferroelectric Memory Applications," J. Electronic Mat., vol. 21, pp. 971-75 (1992).

7. Altaf H. Carim, B. A. Tuttle, A. Doughty and S. L. Martinez, "Microstructure of Solution-Processed Lead Zirconate Titanate (PZT) Thin Films," J. Am. Ceram. Soc., vol. 74 [6], pp. 1455-61 (1991).

8. Yet-Ming Chiang, and Touichi Takagi, "Grain-Boundary Chemistry of Barium Titanate and Strontium Titanate: I, High-Temperature Equilibrium Space Charge," J. Am. Ceram. Soc., vol. 73 [11], pp. 3278-85 (1990).

9. C. K. Barlingay, "Interfacial Reactions, Microstructures, Dopant Compensation, and Electrical Effects in Sol-Gel Pb(Nb,Zr,Sn,Ti)O_3 Films," Ph. D. Thesis, Arizona State University Dec. (1994).

10. M. Raymond, J. Chen, and D. M. Smyth, "Degradation of Ferroelectric Thin Films: A Defect Chemistry Approach," Integrated Ferroelectrics, vol. 5 [1], pp. 73-78 (1994).

11 S. Gurkovich and J. Blum," Preparation of Monolithic Lead Titanate by a Sol Gel Process," in: Ultrastructure Processing of Ceramics, Glasses, and Composites, Edited by L. Hench and D. Ulrich, John Wiley and Sons, New York, 1984.

12. S. K. Dey and R. Zuleeg, "Processing and Parameters of Sol-Gel PZT Thin-Films for GaAs Memory Applications" Ferroelectrics, vol. 112, pp. 309-319 (1990).

13. P. R. Coffman and S. K. Dey, "Structure evolution in the PbO-ZrO_2-TiO_2 sol-gel system: Part I: characterization of prehydrolyzed precursors," J. Sol-Gel Science and Technology, vol. 1, pp. 251-265 (1994).

14. C. K. Kwok and S. B. Desu, "Pyrochlore to perovskite phase transformation in sol-gel derived lead-zirconate-titanate thin films," Appl. Phys. Lett., vol. 60, pp. 1430-36 (1992).

15. R. B.Atkin and R. B. Fulrath, "Point Defects and Sintering of Lead Zirconate-Titanate," J. Am. Ceram. Soc., vol. 54 [5], pp. 265-70 (1971).

16. G. V. Lewis, C. R. A. Catlow, "Defect Studies of Doped and Undoped Barium Titanate Using Computer Simulation Techniques," J. Phys. Chem. Solids, vol. 47 [1], pp. 89-97 (1986).

17. G. R. Fox and S. B. Krupanidhi, "Nonlinear electrical properties of lead-lanthanum-titanate thin films deposited by multi-ion-beam reactive sputtering," J. Appl. Phys. vol. 74, pp. 1949-56 (1993).

18. D. K. Das Gupta and K. Joyner, "On the nature of absorption currents in polyethylene terephthalate (PET)," J. Phys. D: Appl. Phys., vol. 9, pp. 829-38 (1976).

19. J. J. Dih, R. M. Fulrath, "Electrical Conductivity in Lead Zirconate-Titanate Ceramics, "J. Am. Ceram. Soc., vol. 61 [9-10], pp. 448-451 (1978).

20. R. Gerson, H. Jaffe, "Electrical Conductivity in Lead Titanate Zirconate Ceramics," J. Phys. Chem. Solids, vol. 24 [8], pp. 979-84 (1963).

21. M. Takahashi, "Electrical Resistivity of Lead Zirconate Titanate Ceramics containing Impurities,' ibid., vol. 10 [5], pp. 643-51 (1971).

22. T. Ikeda, Y. Tanaka, T. Ayakawa, H. Naoke, "Precipitation of Zirconia Phase in Niobium-Modified Ceramics of Lead Zirconate-Titanate, "Jap. J. Appl. Phys., vol. 3 [10], pp. 581-587 (1964).

Sol-Gel Derived Ferroelectric Films As High Dielectric Constant Materials

G. Teowee, Donnelly Corp., Tucson, AZ 85712
J.M. Boulton and D.R. Uhlmann, Dept. of Materials Science and Engineering, University of Arizona, Tucson, AZ 85721

ABSTRACT

Ferroelectric (FE) films have received considerable attention for a multitude of applications involving FE, dielectric, piezoelectric, pyroelectric, acoustooptic, non-linear optical and electrooptic properties. The present paper discusses the development of sol-gel derived FE films, including a comparison with physically or chemically vapor deposited films, particularly with reference to ultra high density DRAMs. The next generations of ULSI DRAMs will require a substantial enhancement in dielectric constant (up to 1000) which can be provided by FE films such as PLZT, PMN or (Ba,Sr)TiO_3. Materials science issues relevant to the processing of FE films are discussed, as are recent results obtained in our laboratories.

INTRODUCTION

Ferroelectric (FE) films are useful for many applications including SAW devices, high density DRAMS, piezoelectric resonators, electrooptic (EO) switches, non-volatile FE memories, optical data storage, photovoltaics and integrated optics [1]. A wide range of techniques can be used to prepare FE films. These include physical and chemical vapor deposition techniques as well as wet chemical (sol-gel) methods. Not all FE films exhibit similar properties, even for the same nominal composition as differences in deposition techniques and conditions and their effects on film microstructures can produce substantial modifications in properties.

To the extent authorized under the laws of the United States of America, all copyright interests in this publication are the property of The American Ceramic Society. Any duplication, reproduction, or republication of this publication or any part thereof, without the express written consent of The American Ceramic Society or fee paid to the Copyright Clearance Center, is prohibited.

Much of the recent attention on FE films has been directed to the technical implementation of non-volatile FE memory [2] which relies necesssarily on the switching or hysteretic nature of FE films. The market for non-volatile FE memory is expected to increase significantly in the next few years. Ramtron, e.g., has announced a 64K FRAM (ferroelectric random access memory) and similar products are expected from NEC, Matshushita and Motorola. The classic obstacles of fatigue and retention are fast being remedied. The authors envisage a revolution in electronic and electrical devices incorporating high density non-volatile FE memories.

A further interesting area of application area for FE films is in ultra high density DRAMs. Here use is made of the inherently high dielectric constants offered by such films. Notable examples are PZT, $BaTiO_3$ and various relaxor FE's such as lead magnesium niobate (PMN). The critical need for high ε_r materials for the future generations of DRAM's is illustrated in Table, 1 which shows the projected DRAMs, their feature sizes and the dielectrics that will be used.

Table 1. Future DRAM Requirements

Year	**DRAM**	**Feature Size (μm)**	**Cell Area (μm^2)**	**Speed (MHz)**	**Dielectric**
1999	256 Mb	0.25	0.5	400	Ta_2O_5, high ε_r
2002	1024 Mb	0.18	0.2	500	high ε_r
2005	4 Gb	0.12	0.1	600	high ε_r
2008	16 Gb	0.10	< 0.1	700	high ε_r
2011	64 Gb	0.08	<< 0.1	800	high ε_r

Wet chemical processing [3] offers particular promise for synthesizing FE films for many applications [4-6]. The attractive features of such processing include: (1) the solutions employed can be vehicles for incorporating specific active constituents; (2) the highly porous intermediate state can be used to good advantage, e.g. for tailoring ε_r; (3) the ease of changing composition makes available a wide range of dielectric and FE properties; (4) the range of properties between the just deposited gel and final ceramic opens a useful window in mechanical properties which permit procedures such as embossing to be applied to coatings; (5) for multicomponent coatings, problems with vapor phase deposition methods (use of multiple targets, preferential deposition of specific components, etc.) are substantially reduced or eliminated; and (6) the ability to coat large area and non-flat geometries with economic efficiency opens a range of new opportunities.

The present paper will review the recent developments, including materials and processing issues, in the wet chemical processing of FE coatings including the dependence of the film properties on material chemistry, microstructure and processing conditions.

Film Chemistry

The commonly used dielectric material is SiO_2 and the Si-SiO_2 interface has been extensively characterized. SiO_2 is stable, inert, easily prepared and compatible with Si microelectronics processing ; and it has been the workhorse of MOS device technology. Its dielectric constant, ε_r is only 3.8; and while higher capacitance per unit area can be achieved by using thinner layers of SiO_2 there is a practical lower limit to the thickness of the films before direct tunnelling currents or processability considerations become technical obstacles. Novels ways to enhance the effective capacitance based on SiO_2 include trench capacitors, stack capacitors and complex 3-D capacitive stuctures (obtaining high aspect ratio device dimensions is crucial).

Higher ε_r materials such as Si_3N_4 (in oxide - nitride - -oxide stacks) or Ta_2O_5 - with ε_r 's of 5 and 25 respectively - came into use for high density DRAM'S (e.g. up to 4Mb). Other candidate materials include Y_2O_3, HfO_2 and TiO_2 (see Table 2). While these materials exhibit relatively high values of ε_r, they suffer from high leakage currents and require high processing temperatures for complete densification. These oxides are deposited as paraelectric single-component films. For still higher density chips incorporating planar configurations with smaller submicron feature size, higher ε_r materials are required with values up to 1000 for a small film thickness. For 256Mb and higher density DRAM chips, a charge storage density of at least 14 μC/cm2 is required [7].

Table 2: Dielectric Constants of Dielectric Films.

Films	Dielectric Constant
SiO_2	3.8
Si_3N_4	7
Ta_2O_5	15 - 25
Y_2O_3	15
ZrO_2	15-30
TiO_2	100
HfO_2	45 - 150

High quality FE films , which are usually multicomponent in nature, must have high purity, homogeneity and stoichiometry (all chemical issues). With regards to these issues, sol-gel FE films are expected to yield superior dielectric and electrical properties. Such films used in high ε_r applications are preferably paraelectric; the judicious choice of dopants is used to shift the Curie temperatures, T_c's, to room temperature. Doping of PZT compositions has been demonstrated in the literature. Numerous dopants have been investigated, most notably the incorporation of La has been widely studied to yield materials with a wide range of desirable properties. The incorporation of La into PZT can lead to A (i.e. Pb) or B (i.e. Ti) site vacancies i.e. $Pb_{1-3x/2}\,La_x\,TiO_3$ or $Pb_{1-x}\,La_x\,Ti_{1-x/4}\,O_3$ [8]. The Pb content can vary over a wide range fora given La content. Excess PbO is often incorporated in PZT films to compensate for eventual PbO loss due to volatilization and diffusion into underlying substratesupon heat treatment [9]. Such excess PbO tends to give rise to an optical absorption peak about 440 nm, often resulting in a yellowish color.

Microstructural Issues

While films of FE compositions can be amorphous, single crystalline or polycrystalline depending on the precursors and processing conditions , those which exhibit FE properties must necessarily be crystalline. The lowest dielectric losses and leakage currents are obtained with amorphous or single crystalline films as grain boundaries in polycrystalline films are regions of compositional and structural discontinuities which produce undesirable variations in local electronic structue.

Since crystalline FE's are necessarily anisotropic, the degree of preferred orientation with respect to the electrodes can strongly influence the relative dielectric constant. For example, the dielectric constant of BaTiO3 is higher along the a direction than along the c-direction. Crystalline orientation of films is achieved via substrate epitaxial effects, the deposition process itself or by unidirectional shrinkage and densification/crystallization during post-deposition processing. The potential for manipulating film orientation by choosing the appropriate substrate to effect the different thermal expansion offers considerable promise but remains relatively unexplored. Single crystal or epitaxial films have been studied, usually formed on Pt on MgO or $SrTiO_3$. Such single crystalline films are difficult to obtain and usually involve expensive substrates and elaborate processing conditions. Another aspect of film orientation involves the domain structure in films exhibiting FE characteristics, which in most cases are constrained by crystallographic directions. Poling of the films can be performed to influence the orientation of the film during or after crystallization. The poling direction can be along or perpendicular to the substrate surface resulting in the alignment of the polarization vectors in their respective directions and usually leading to significant enhancements in dielectric properties.

Deleterious film-substrate interactions are typically more extensive in polycrystalline films (due to preferential segregation and favorable diffusion along grain boundaries) than in amorphous or single crystal films. The presence of interfacial layers with low ε_r's may lower the overall ε_r of the films and may explain the dependence of ε_r on film thickness. Si is well documented to react with PZT or PLZT films forming complex lead silicate species - thus the direct deposition of PZT on Si is rarely done without the use of intermediate barrier layers. Much effort has been directed to exploring contacts or electrodes for PZT and other FE's. The PZT- substrate interface significantly affects the FE properties and especially the fatigue behaviour. The bottom electrode or substrate usually consists of a noble metal such as Pt. Oxide electrodes such as RuO_2 and YBCO superconductor have attracted much recent attention , and have been reported to yield fatigue-free devcies up to 10^{12} cycles. The present authors have explored the effect of top metal electrodes on the dielectric and FE properties of sol-gel derived PZT capacitors. It was found that the work function alone could not fully explain the observed dependence of the dielectric properties on the choice of electrodes, and that the reactivity of the electrode and surface states at the PZT-electrode junctions had a considerable impact on the capacitor characteristics [12a].

Important features of polycrystalline FE films, which distinguishes them from conventional paraelectric films, are the grain size effect - i.e., the disappearance of ferroelectricity (and appearance of paraelectric behaviour) below a critical grain size below some range of thickness - and the film thickness effect (where ε_r decreases with decreasing thickness). Larger grain size films usually enhance FE and EO properties, while smaller grain size films offer lower optical attenuation and lower processing temperatures. The effect of grain size may also explain why the values of ε_r tend to be smaller in thinner films than in thicker films since the grain size in thicker films is larger. Another important consequence of the size effect, and the disappearance of ferroelectricity below a certain grain size , is the opportunity for designing paraelectric films for DRAM applications in films with nominally FE compositions. With the low processing temperatures inherent in sol-gel processing, this becomes a real opportunity. For most FEs, the decrease of ε_r in thinner films pose a scaling problem in ULSI devices since the value of ε_r will be a strong function of film thickness below about 2000 - 3000 A . Besides a smaller grain size such observations have been attributed to the presence of paraelectric low ε_r layers at the PZT-electrode junctions [12b]. The film thickness effect is well likely to dominate design considerations for future ULSI DRAM's since the films are thinner as the density becomes higher and for the same given voltage, the breakdown strengths of the films will be crucial.

Phase Assemblage Issues

Phase assemblages have an important effect on the dielectric properties of the films. For crystalline FE films, the amount of the desired FE phase (usually perovskite phase) should be maximized, and the porosity and amorphous/pyrochlore phase sholud be minimized for optimal FE properties. The crystallization behaviours of films is often quite complex and involves the competition between organic burnout, densification of the gel matrix and homogenous/heterogenous nucleation, sometimes accompanied by the undesired formation of paraelectric metastable phases. For epitaxial growth of films, heterogenous nucleation should predominate and homogenous nucleation suppressed over some temperature range. For example, in PZT films, a metastable pyrochlore phase usually forms prior to the appearance of the FE perovskite phase; this pyrochlore phase is paraelectric and tends to degrade the overall FE and dielectric characteristics of the films. Additionally, the crystallization behaviour can be substantially different in thin films compared to bulk materials, due to the presence of substrates and the much higher surface-to-volume ratios in the films. Porosity can be viewed as an extraneous phase and is present to some degree in almost all as-prepared films. It can be discrete but is often interconnected. It is desirable to minimize porosity since it can contribute to pinhole formation. In order to obtain fully dense crystallized films, it is often desirable to achieve fully dense films after final organic burnout prior to crystallization.

Processing Issues

One of the most important processing issues is the thermal limitation on processing the entire wafer containing the FE and semiconductor devices together. For most Si-based on-wafer devices, the upper temperature limit is about 500C (at the most) in order not to perturb underlying junction or oxide profiles. Thus any FE films which are deposited on Si-devices during the last steps of device patterning must be processed at temperatures less than 500C; alternatively FE films which require considerably higher processing temperatures must be deposited (before the Si-based devices are fabricated). Rapid thermal annealing can also be used to good advantage in such processing. Such rapid thermal annealing is beneficial in obtaining the desired perovskite phase with small grain size.

Dry etching of Pb-based compounds can be difficult and presents enviromental concerns. Classical dry etching of FE materials involves mixed gases of Cl_2 and SF_6 and is used to define patterns in $SrTiO_3$ and PZT [13]. The multicomponent nature of most FE films exacerbates the difficulties associated with this process due to differential volatilities of the various gaseous by-products. For the most part, dry etching of PZT or other Pb-based FE films is still an area of unproven technology and requires critical input from engineers and device technologists.

Mechanical Issues

From a mechanical perspective, the films should exhibit high values of hardness(for scratch resistance), and low (if tensile) or controlled (if compressive) stresses within the films as well as good adhesion to the substrates to avoid cracking or delamination during processing and use. Since the FE films are non-centrosymmetric, stresses within the films can notably affect the domain structure, dielectric and FE properties, and phase transitions of the films. Stresses may arise from the deposition process itself or from differential thermal expansions of the films and substrate. If the stresses are non-uniform through the films, the variation in ε_r may be substantial, resulting in reproducibility problems.

FE Materials

For high ε_r materials used in microelectronic applications, such as DRAMs, the films should ideally consist of linear dielectrics (to avoid hysteretic effects seen in FE films). Such films should exhibit high values of ε_r (500 - 1000), low values of leakage current (preferably < 1×10^{-5} A/cm^2 at 3V) and low dielectric losses, high dielectric breakdown strengths, low switching times (< 1ns) and low fatigue (up to 10^{12} - 10^{15} cycles). Another important reason for desiring paraelectric films, besides the avoidance of hysteresis, is the strong dispersion of ε_r at high frequencies due to domain and domain wall relaxation.

The rf (1kHz - 1 Mhz) dielectric constants of some important FE and paraelectric films are shown in Table 3. PMN:PT and PLZT films are seen to exhibit the highest dielectric constants (4000 - 6000). PZT 53/47 films can exhibit ε_r values as high as 3000 depending on the voltage pulse used; but these films are ferroelectric and are hysteretic. Paraelectric PLT 28 films, with ε_r of 1400 [28] have been proposed as a dielectric material in ultrahigh density DRAMs, while sol-gel derived

$(Ba,Sr)TiO_3$ films have been commercialized in Matsushita DRAM chips. Other FEs such as $LiNbO_3$, KNbO3 or KTN (potassium tantalum niobate) exhibit low values of ε_r. Additionally the presence of alkali ions is anathema in CMOS microelectronic processing due to charge migration and trapping at the interfaces, leading to device instability.

An important issue in comparing these values of ε_r is the non-linear dielectric effect observed in FE's, i.e. the phenomenon of dielectric saturation manisfested by the levelling off of polarization or ε_r at high electric fields. Even in paraelectric PLT films which are polar, the same phenomenon is observed. Linear oxide dielectrics such as SiO_2 exhibited ε_r's which are independent of field up to breakdown voltages. Thus the values of ε_r in FE or paraelectric films are usually lower under applied dc bias, which impacts on the capacitance measured under high applied fields. For example, ε_r decreases from 1400 at low applied fields to 850 at an applied field of 200 kV/cm in PLT 28 films [28]. Thus for viable comparisons of ε_r, the capactiors should nominally be subjected to the same probing voltages (3 - 5 V) present in typical DRAM circuitries.

The effects of fatigue are well documented in FE PZT films [e.g. 32] and involves the polarization reversals of the domains in these films. For DRAM operation, the effect of unipolar switching has not been extensively investigated. Preliminary results obtained in the authors' laboratories have shown that there may be some "fatigue" effects, manisfested by a lowering of ε_r with unipolar pulses. It is hoped that more studies will be forthcoming of this exceedingly important reliability issue.

Table 3: Dielectric constants, ε_r's, of various solgel and non-solgel FE and paraelectric films

FE Film	Er	Method	Reference
PMN	1800	Sol-gel	14
PMN	1000-1250	Sol-gel	15
PMN:PT	2000	Laser Ablation	16
PMN:PT	2200	Sol-gel	17
PMN:PT	2000	Sol-gel	18
PMN:PT	2900	Sol-gel	19
PMN:PT	6000	Sol-gel	20
PLZT 7/65/35	2100	Sputtered	21
PLZT 8/65/35	1500	Solgel	22
PLZT 9/65/35	1000	Sputtered	23
PLZT 9/65/35	2600 - 6600	Sol-gel	This work
PLZT 9/65/35	4000	Sol-gel	24
PZT 53/47	1700	Sol-gel	This work
PZT 50/50	1200	Sol-gel	25
PZT 35/65	1200	Sputtered	26
PZT 55/45	1300	Sol-gel	27
PLT 28	1400	Sol-gel	28
PLT 28	1200	Sol-gel	This work
PLT28	2000	MOCVD	29
$(Ba,Sr)TiO_3$	500	Sol-gel	30
$(Ba,Sr)TiO_3$	450	Coevaporation	31

The following ferroelectrics are being seriously considered as candidates for high ε_r materials:

1. PLZT

The PLZT family includes lead titanate ($PbTiO_3$), lead zirconate titanate (PZT), lead lanthanum titanate (PLT) and lead lanthanum zirconate titanate (PLZT). By judicious choice of the Pb, Zr, Ti and La stoichiometries, FE, anti-FE and paraelectric behaviours can be obtained in PLZT. In combination with various dopants, they constitute the most versatile class of perovskite ferroelectrics, and are extensively employed in applications ranging from transducers and optical modulators to electrooptic shutters. PLZT films have been prepared by numerous techniques including sputtering, laser ablation, MOCVD, evaporation and solgel methods. Films derived from solgel methods typically exhibit superior properties to films prepared by other techniques. This attests to the high degrees of stochiometry and homogeneity achieved in the sol-gel films.

PLT films with La content > 28 mole % are paraelectric at room temperature; and ε_r increases dramatically with La content since the Curie temperatures are shifted toward room temperature with increasing La content. The dielectric and FE properties of rare-earth doped PbTiO3 films have also been explored by the present authors, yielding ε_r of up to 600 [33]. PLT 28 films exhibited values of ε_r up to 2000 [29] and are attractive for DRAM applications. However, being non-ferroelectric, PLT 28 is polar and hence shows some degree of dielectric polarization - i.e., the dielectric proeprties are affected by high applied fields. For example ε_r of PLT 28 is fairly constant at 1400 up to 40 kV/cm, but declines to 850 at 200kV/cm [28]. PLZT films exhibit fairly high values of ε_r; PLZT 9/65/35 films have been reported to yield ε_r of 4000 [24]; and data on such films synthesized in our laboratories indicated that ε_r as high as 6600 can be readily obtained. While these values are attractive,, they are still substantially below values for bulk ceramic PLZT, likely reflecting incomplete densification (remanent porosity) or microstuctural (grain size) differences.It should be noted, however, that the PLZT 9/65/35 and related PLZT x/65/35 compositions tend to be more characteristic of relaxor FE's rather than normal ferroelectrics

2. Barium Strontium Titanate - $(Ba,Sr)TiO_3$

$BaTiO_3$ is perhaps the most studied FE material. The room temperature value of ε_r can range from 2000 to 10,000 depending on the orientation of the single crystal and the grain size in polycrystalline ceramics. Its close cousin, $SrTiO_3$, is cubic and hence nominally paraelectric at room temperature. The value of ε_r of $SrTiO_3$ is only about 250. $BaTiO_3$ is strongly anisotropic and as in the case with $PbTiO_3$-based materials, it can be used for a variety of applications. The recent attention on Ba ans Sr titanates, including solid solutions between them, has been focussed on application as high ε_r materials in ultrahigh density DRAM's. $(Ba,Sr)TiO_3$ films with various Ba/Sr stochiometries have been obtained using sputtering, MOCVD and sol-gel processing. For example, $BaTiO_3$:SrTiO3 films of 70:30 composition and 148 nm thick exhibited a value of 500 for ε_r and leakage current of 1.2×10^{-8} A/cm^2 at 4V [30]. The value of ε_r for a corresponding bulk ceramic is about 2500 [34]. While the ε_r's are notably lower than those of PZT or PLZT, at high frequncies (GHz), the ε_r of PZT decreases to about 40 while in $(Ba,Sr)TiO_3$ films, there is no appreciable decrease up to 3 - 4 GHz [35].

3. Relaxor Ferroelectrics

Relaxor FE's (also known as diffuse phase transition FEs) have a broad Curie maximum rather than the sharp and distinct T_c's of normal FEs. This effect is suggested to arise from the micro-inhomogeneous microstructure which consists of paraelectric and ferroelectric regions. Each region has its own T_c. Lattice and/or cation disorder is responsible for the compositional heterogeneity. PMN ($Pb(Mg_{1/3}Nb_{2/3})O_3$) is the most widely studied relaxor [36]. Its high peak value of dielectric constant, ε_r, of >15,000, makes it an attractive candidate for ceramic multilayer capacitors and possibly thin films.

Analogous to PZT, the morphotropic phase boundary in the solid solution of relaxors and normal FE's (e.g. PMN - PT at 65 mole% PMN - 35 mole% PT) provides the highest values of dielectric constant, electrostrictive coefficent and piezoelectric properties. There are many FEs which exhibit morphotropic phase boundaries including SBN (strontium barium niobate), PZN (lead zinc niobate) - PT and KTN (potassium tanalum niobate), which offer rich possibilites for obtaining high values of ε_r.

Several wet chemical routes to PMN films have been explored[14,15]. The results are summarized in Table 3 above. In the all-alkoxide route, pure perovskite was formed at 800C on single-crystal MgO; however, on Pt/Si, a small amount of pyrochlore remained. Oriented films could be prepared on Pt/Si, with high initial heating rates (~$1x10^5$ °C/h) favoring (100) over (111) orientation [37]. For the 2-methoxyethanol route, (100) oriented films were obtained upon fast-firing at 800C for 5 min [15]. The values of ε_r of PMN films (1800) are low compared with those of bulk polycrystalline and single-crystal material. The higher ε_r of (100) compared with (111) oriented films is expected, since in the former case the measurement is perpendicular to the direction of the spontaneous polarization. The peak of the distribution of Curie temperatures of PMN is approximately -12°C. It can be shifted to higher temperature by forming a solid solution with the normal perovskite FE, $PbTiO_3$ (PT). The PMN-PT system shows a morphotropic phase boundary (MPB) in the vicinity of $x \approx 0.35$ for $(1-x)Pb(Mg_{1/3}Nb_{2/3})O_3 \bullet xPbTiO_3$ [38].

The dielectric properties of the PMN:PT films are summarized in Table 3 above. As found for PMN, PMN-PT films exhibit values of ε_r which are much smaller those of bulk samples (~19,500 for bulk ceramic prepared from wet chemically derived powder, sintered at 950C for 2 h). Results in our laboratories indicate that some relaxor PLZT films exhibit ε_r up to 6600, one of the highest recorded ε_r of any sol-gel-derived FE film.

A model of the Pt-FE-Pt capacitor has been proposed by the authors [39] which was based on totally depleted back-to-back Schottky barriers rather than partially depleted films with a neutral region sandwiched by the two depletion regions. This model proposed the presence of a built-in field existing within the FE films, whose magnitude and sign are dependent on the film thickness. This built-in field smears out the FE properties of relaxor FE films (e.g. PMN), which are much more sensitive to electric fields than are films of normal FE's such as PZT or $BaTiO_3$.
The observation of substantially smaller values of dielectric constants for relaxor thin films (by an order magnitude compared to bulk ceramics) has remained a puzzle - for example the dielectric constants of PMN and PMN:PT films were reported as 1800 and 2200 respectively, while the values for the corresponding bulk ceramics were 15,000 and 19,500, respectively. For relaxor FE films thicker than the depletion width, the overall capacitance (and hence ε_r) will be dominated by the depletion capacitance unless the films are thick enough that the majority of the voltage drop takes place in the bulk or neutral portion of the film rather than in the depletion zones. As a first approximation, the transition to bulk behaviour for relaxor FE films should not be expected unless the films are at least 5 - 10 um thick. For thinner films, with greater likelihood of being fully depleted, the effect of such built-in fields is greater than in thicker films. Hence it can be surmised ,that following the totally depleted FE capacitor model, that relaxor films are not suitable for applications in ultrahigh density DRAMs since near-bulk values cannot be achieved (since in the next generation of DRAMs the thickness of the films are limited to < 1000 A) and also these relaxor FE's are characterized by high dispersion.

CONCLUSIONS

Wet chemical methods offer promise for forming FE thin films of a range of materials for electrical applications. Potential advantages of such methods include low processing temperatures, avoidance of columnar structures, ease of introducing dopants and the potential ease of forming chemically homogeneous and stochiometric multicomponent films. A wide range of oxide FE films have been prepared by wet chemical methods. Such methods typically yield films with superior dielectric and FE film properties in many aspects compared with physically deposited films. While normal perovskite FE films exhibit properties similar to those of bulk materials, thin films of relaxor FEs consistently show inferior - often greatly inferior - properties compared with those of bulk ceramics.

The greatest opportunities in sol-gel processing of FE films for high ε_r materials lie in the mastery of chemistry and processing to control molecular architecture and microstructure. New materials with novel combinations of properties can be created. To capitalize on these opportunities, it will be necessary to undertake more systematic studies of chemical and structural changes during processing; of the coating process itself; of post-deposition processes; and most important, of the microstructure of thin films and the phenomena which are important in tailoring and controlling their microstructures (e.g. FE film - substrate interfacial reactions). While the current efforts in tailoring materials used in DRAMs illustrate the collaboration between sol-gel chemists and electrical engineers, more effective collaboration needs to be established between chemists/ceramists who are knowledgeable about the intricacies of wet chemical processing, and device techologists who are knowledgeable about the chemistries/structures/properties of coatings which can have major impact upon technology in hybrid packaging. The present authors are highly optimistic about the future of wet chemically-derived FE films for such applications.

ACKNOWLEDGEMENTS

Financial support for the present work has been provided by the Air Force Office of Scientific research and by Donnelly Corporation. This support is gratefully acknowledged.

REFERENCES

1. M.H.Francombe, Phys. Thin Films, 17, 225 (1993).
2. W.A.Geideman, IEEE Trans. Ultra. Ferroelectric Freq. Control, 38, 704 (1991).
3. G.P.Rajendran and D.R.Uhlmann, Ultrastructure Processing of Advanced Ceramics (Eds. J.D.MacKenzie and D.R.Ulrich , Wiley 1988), 241.
4. D.R.Uhlmann, G.Teowee, J.M.Boulton, S.Motakef and S.C.Lee, J.Non-Cryst. Solids, 147 & 148, 409 (1992).
5. D.R.Ulhmann, G.Teowee and J.M.Boulton, Glastech. Ber. Glass Sc. Technol., 67C, 55-62 (1994).
6. G.Teowee, J.M.Boulton, S.Motakef, D.R.Uhlmann, B.J.J.Zelinski, R.Zanoni and M.Moon, SPIE, 1758, 236 (1992).
7. J.Lee, V.Chikarmane, C.Sudhama, J.Kim and Al Tasch, Proc. 4th Int'l Symp. Integr. Ferroelectrics, 298 (1992).
8. C.E.Land, P.D.Thacher and G.H.Haertling, Appl. Solid State Sci., 4, 137 (1974)
9. G.Teowee, J.M.Boulton and D.R.Uhlmann, MRS Symp. Proc, 271, 345 (1992).
10. S.D.Bernstein, T.Y.Wong, Y.Kisner ans R.W.Tustin, J. Mater. Res., 8, 12 (1993).
11. R.Ramesh, W.K.Chan, B.Wilkens, T.Sands, J.M.Tarascon and V.G.Keramidas, Integr. Ferroelectrics, 1, 1 (1992).

12a. G.Teowee, C.D.Baertlein, E.A.Kneer, J.M.Boulton and D.R.Uhlmann, Cer.Trans., 43, 401 (1994).

12b. . K.Amanuma, T.Mori, T.Hase, T.Sakuma, A.Ochi and Y.Miyasaka, Jpn. J.Appl. Phys., 32, 4150 (1993).

13. H.Aoki, T.Hashimoto, E.Ikawa, T.Kikkawa, K.Takeuchi, Y.Yamaguchi, T.Sakuma and Y.Miyasaka, Jpn. J. Appl. Phys., 32, 376 (1993).
14. K.Okuwada, M.Inai and K.Kakuno, Jpn. J. Appl. Phys., 28, L1271 (1989).
15. L.F.Francis, Y.J.Oh and D.A.Payne, J. Mater. Sci., 25, 5007 (1990).

16 C.Tantigate, A.Safari and J.Lee, presented at 6th Int'l Symp. Integr. Ferroelectrics, Monterey, March 1994
17. L.F.Francis and D.A.Payne, MRS Symp. Proc., 200, 173 (1990).
18. L.F.Francis and D.A.Payne, Proc.7th IEEE Int'l Symp. Appl. Ferroelectrics, 263 (1990).
19. K.R.Udayakumar, J.Chen, P.J.Schuele, L.E.Cross, V.Kumar and S.B.Krupanidhi, Appl. Phys. Lett., 60, 1187 (1992).
20. J.Chen, Q.Zhang and L.E.Cross, MRS Symp. Proc., 310, 47 (1993).
21. T.Nakagawa, T. Yamaguchi, T.Usuki, Y.Matsui, M.Okuyama and Y.Hamakawa, Jpn. J. Appl. Phys., 18, 897 (1979).
22. R.W.Vest and J.Xu, Ferroelectrics, 93, 21 (1989).
23. H.Adachi, T.Mitsuyu, O.Yamayaki and K.Wasa, J. Appl. Phys., 60, 736 (1986).
24. D.E.Dausch and G.H.haertling, Proc. 8th IEEE Int'l Symp. Appl. Ferroelectrics, 297 (1992).
25. R.W.Schwartz, D.Dimos, S.J.Lockwood and V.M.Torres, MRS Symp. Proc., 310, 59 (1993).
26. J.Kim, S.Sudharma, R.Khamaukar and J.Lee, MRS Symp. Proc., 310, 473 (1993).
27. M.Klee, R.Eusemann, R.Waser, W. Brand and H.Van Hal, J.Appl. Phys., 72, 1566 (1992).
28. S.K.Dey and J.Lee, IEEE Trans. Elec. Dev., 39, 1607 (1992).
29. H.Adachi, T.Kawaguchi, K.Setsune. K.Ohji and K.Wasa, Appl. Phys, Lett., 42, 867 (1983).
30. M.Azuma, M.Scott, E.Fujii, T.Otsuki, G.Kano and C.A.Paz de Araujo, Proc. 4th Int'l Symp. Integr. Ferroelectrics, 109 (1992).
31. H.Yamaguchi, S.Matsubara, K.Takemura and Y.Miyasaka, Proc. 4th Int'l Symp. Integr. Ferroelectrics, 285 (1992).
32. G.Teowee, E.L.Quackenbush, C.D.Baertlein, J.M.Boulton and D.R.Uhlmann, to be published in Proc. 9th IEEE Int'l Symp. Appl. Ferroelectrics, 1994.
33. G.Teowee, C.D.Baertlein, S.A.Schlegel, J.M.Boulton and D.R.Uhlmann MRS Symp. Proc., 346, 291 (1994).

34. U. Syamaprasad, Mater. Lett., 7, 56 (1988).
35. J.F.Scott, M.Azuma, E.Fujii, T.Otsuki. G.Kano, M.C.Scott, C.A. Paz de Araujo, L.D.McMillan and T.Roberts, Proc. 8th IEEE Int'l Symp. Appl. Ferroelectrics, 356 (1992).
36. S.L.Swartz and T.R.Shrout, Mater. Res. Bull., 17, 1245 (1982).
37. K.Okuwada, S.Nakamura, M.Imai and K.Kakuno, Jpn. J. Appl. Phys., 29, 1153 (1990).
38. S.W.Choi, T.R.Shrout, S.J.Jang and A.S.Bhalla, Ferroelectrics, 100, 29 (1989).
39. G.Teowee and D.R.Uhlmann, Cer. Trans., 43, 255 (1994).

MINIMIZATION OF FATIGUE IN FERROELECTRIC FILMS

Seshu B. Desu
Department of Materials Science and Engineering, Virginia Tech
Blacksburg, VA 24061

ABSTRACT

On the basis of a quantitative fatigue model, various methods to minimize fatigue in oxide ferroelectric films are discussed. The model attributes fatigue to domain pinning by space charge that is caused by oxygen vacancy entrapment at the interfaces (e.g., electrode-ferroelectric, domain boundaries etc.). Based on this model, it has been demonstrated that fatigue can be minimized by : (a) using conductive oxide electrodes (e.g., RuO_2) that prevent space charge formation at the interfaces; (b) by addition of donor dopants (e.g., Nb, La in $Pb(Zr_xTi_{1-x})O_3$) that reduce the oxygen vacancy concentration and (c) by utilizing ferroelectric materials (e.g., $SrBi_2(Ta_xNb_{1-x})_2O_9$) that have a low intrinsic defect concentration.

INTRODUCTION

Over the last few decades, there is an increasing interest in ferroelectric materials for nonvolatile ferroelectric memory (FRAM) applications [1-3]. In principle, the memory application of these materials is based on the hysteresis behavior of polarization with applied electric field as shown in Fig. 1. At zero applied field, there are two equally stable states of polarization, $\pm P_r$ (remnant polarization). Integration of submicron thin film ferroelectric capacitors, for e.g., lead zirconate titanate (PZT), into the existing silicon based transistor technology can provide a transistor-capacitor based ferroelectric memory cell. Either of the two states of polarization could be encoded as a '1' or a '0' in the memory device and since no external field is required to maintain these states, the memory device is nonvolatile.

To the extent authorized under the laws of the United States of America, all copyright interests in this publication are the property of The American Ceramic Society. Any duplication, reproduction, or republication of this publication or any part thereof, without the express written consent of The American Ceramic Society or fee paid to the Copyright Clearance Center, is prohibited.

Ferroelectric memories are most often designed with destructive electrical read out. To read the stored information, an electric field greater than the threshold field (E_c, see Fig. 1) is applied to each memory cell. In memory elements where the polarization is opposite to applied field direction (for e.g., state '1' for a positive applied field in Fig. 1), there is a large displacement current accompanying the polarization reversal while only a displacement current occurs in elements having a polarization assisting the field (e.g., state '0' for a positive field). The state of each memory element is therefore read by measuring the displacement current corresponding to each voltage pulse. The difference between the two displacements currents corresponds to the net amount of switched charge for each polarization reversal [4]. Evidently, this kind of an electrical read out involves several polarization reversals in the ferroelectric films over the device lifetime.

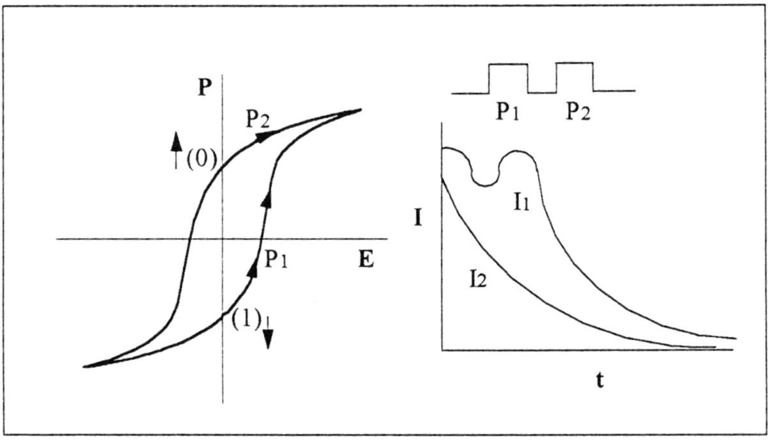

Fig. 1: Typical hysteresis loop and displacement currents of a ferroelectric digital memory capacitor (Ref. 4).

In some of the more popular metal-ferroelectric-metal capacitor systems such as Pt/PZT/Pt, a progressive decrease in the amount of switched charge is commonly observed with increasing polarization reversals. Eventually, after a large number of read/erase/write operations, the switchable polarization decreases to a value that is too low to distinguish between the states, thereby rendering the memory cells ineffective. Additionally, the threshold field required to switch the state of the device also increases with increasing polarization reversals. This degradation phenomenon is known as fatigue and is one of the more profound problems hampering the development of commercial FRAM devices.

Fatigue in ferroelectric devices has been studied for several years and several mechanisms have been proposed to explain this degradation phenomena [5-8]. A theoretical expression for polarization decrease was developed by Karan [5] based on space charge entrapment inside the ferroelectric that results in an increase in the interfacial resistance and decrease in the interfacial capacitance. Zheludev [6] proposed a planar model based on the formation of an inactive surface by space charge at the electrode-ferroelectric interfaces to produce a smaller amount of switched charge. Kudzin et al [7] ascribed fatigue to the accumulation and resorption of space charge at the domain boundaries. The domain wall loses its capacity to move and the resultant value of the switching charge diminishes. Duiker et al [8] simulated fatigue by assuming dendritic growth of oxygen-deficient filaments. In this paper, the problem of fatigue is discussed on the basis of a recent quantitative model that has been developed by the author [9-12]. Using this model, various methods to improve fatigue are then demonstrated.

FATIGUE MODEL

As mentioned previously, fatigue in ferroelectric capacitors is manifest as a loss of switchable polarization and increase in coercive field with increasing number of polarization reversals. Additionally, it is commonly observed that the polarization loss in ferroelectrics such as PZT and KNO_3 can be recovered partly or fully either by applying a voltage pulse higher than the nominal switching voltage [1] or by heating (and subsequently cooling) the ferroelectric material above its Curie temperature. Considering these effects, we have proposed a quantitative fatigue model for Pt/PZT/Pt capacitors in which defect movement under alternating pulses and defect entrapment at trapping sites such as electrode-ferroelectric interfaces, domain boundaries and grain boundaries has been cited as the source for fatigue [9-12]. These point defects could either be extrinsic or intrinsic. Extrinsic defects are impurities (e.g., acceptor atoms in PZT) that are incorporated in the films during processing and can be controlled by controlling the processing environment. Intrinsic defects may include both defects that maintain the stoichiometry (Schottky) and defects that alter the stoichiometry in the materials. Schottky defects in perovskite (ABO_3) ferroelectrics may be represented by a quasi-chemical reaction (in Kroger-Vink notation) as :

$$A_A + B_B + 3O_O \rightarrow V_A'' + V_B'' + 3V_O^{oo} + A_s + B_s + 3O_s \qquad \text{---------} (1)$$

where A_A, B_B, $3O_O$ represent respectively occupied A, B and O sites; V_A'', V_B'', V_O^{oo} represent vacancies of A, B and O atoms; and A_s, B_s, $3O_s$ are the respective Schottky defects. A typical example of defects that alter the stoichiometry are vacancies that are formed due to the vaporization of one or more volatile components in the multicomponent oxide materials. In the case of PZT, for example, a processing temperature of at least 600 °C is required to form the ferroelectric phase. However, the PbO component begins volatilizing at temperatures as low as 550 °C, resulting in the formation of oxygen and lead vacancies as shown below [13]:

$$Pb_{Pb} + Ti_{Ti} + Zr_{Zr} + 3O_O \rightarrow xPbO + xV_{Pb}'' + xV_O^{oo} + (1-x)\,Pb_{Pb} + Ti_{Ti} + Zr_{Zr} + (1-x)\,O_O \qquad \text{----------- (2)}$$

When an a.c. field is applied to a ferroelectric capacitor, the ferroelectric material is polarized and an internal field is created by this polarization. The net local field is then expressed as [9-12] :

$$E_L = E_A - E_i \qquad \text{---------- (3)}$$

where E_A is the applied field and E_i is the internal field. Although the externally applied field is symmetric, the net local field is asymmetric as a result of the internal field contribution. Under this asymmetric local field, there is an effective one-directional movement of the mobile defects (mainly oxygen vacancies as shown in eqns. (1) and (2)) towards the electrode-ferroelectric interfaces. Additionally, the defect segregation to the interfaces with increasing polarization reversals could also be induced by the domain sweeping phenomena. In general, in polycrystalline thin films, the grain growth is usually continuous form the bottom interface to the top surface of the films. In ferroelectric thin films, each of these grains commonly constitutes a single domain (regions containing dipoles oriented in the same direction). In other words, the grain boundaries and the domain boundaries are usually coincident with the electro-ferroelectric interfaces. The large interfacial energy at these boundaries/interfaces provides low potential energy sites for the defects and as a result causes their entrapment. Defect entrapment at the interfaces results in the development of space charge in these regions which in turn causes the domains to be pinned. The reduction in switchable polarization and increase in coercive field with increasing polarization reversals is a direct result of the domain pinning by the space charge.

There is sufficient experimental evidence to indicate that accumulation of oxygen vacancies at the electrode-ferroelectric interfaces has a direct effect on the fatigue properties of the ferroelectric films. Scott et al [14] have noticed a decrease in the oxygen concentration (along with a loss of polarization) near the electrodes after fatigue cycling in Pt/PZT/Pt capacitors indicating increased accumulation of oxygen vacancies near the electrode-ferroelectric interface. Kwok and Desu [15] created near-surface oxygen depletion regions of different thickness in PZT films by processing the films under different partial pressures of oxygen and studied the role of oxygen vacancies on the ferroelectric and fatigue properties of the films. Their results are summarized in Table 1. The films processed under a low partial pressure of oxygen in the processing ambient (i.e., higher oxygen vacancy concentration in the films) showed a lower P_r and a higher E_c. Additionally, these films showed a relatively higher rate of fatigue for a given number of cycles ($\Delta P/P$ for a given N). Lee and Desu [16] have observed an increase in the interfacial capacitance (with no change in the bulk permitivity) with increasing fatigue cycles in Pt/PZT/Pt capacitors. This observation has been attributed to the build up of space charge in the interface regions.

PO_2 (atm.)	P_r ($\mu C/cm^2$)	E_c (kV/cm)	$\Delta P/P$ % (after 10^6 cycles)
10^{-2}	22.5	80	34
2×10^{-1}	30	80	22
1	30	60	21

Table 1 : Role of oxygen vacancies on hysteresis and fatigue properties of PZT films (Kwok and Desu, 1992, Ref. 15)

To develop a quantitative expression for fatigue, the polarization loss due to defect entrapment at the interfaces was assumed to follow the relationship [9-12]:

$$dP/dN = -\nu P \qquad \text{----------- (4)}$$

where P is the polarization with defect entrapment N, and ν is the polarization loss constant related to lattice structures. The number of defect trapped at the interface, at a certain test cycling n, was described by :

$$dN/dn = \Delta j \, \Delta t \qquad \text{----------- (5)}$$

where N is the total defect number trapped at the interface at n, Δj is the net defect flux density between n and n + 1 test cycles, and Δt is the applied pulse width. The internal field decreases as vacancies are trapped at the interfaces. The decrease of internal field difference is proportional to the number of vacancies arriving at the interface at a certain test time and coefficient of dielectric viscosity (β) of the material.

$$d\Delta E_i/dN = -\gamma\beta\Delta j = -\xi\Delta Ei \qquad \text{------------ (6)}$$

where γ is a constant and ξ is related to the coefficient of dielectric viscosity and activation energy for defect movement. Based on equations (4), (5) and (6), a fatigue equation was derived as :

$$P = P_o (An + 1)^{-m} \qquad \text{------------ (7)}$$

where P_o is the initial polarization, P is the polarization after test cycles n, A is defined as the piling constant and m is defined as the decay constant. A and m are voltage and temperature dependent constants related to the dielectric viscosity, defect velocity, defect concentration, activation energy for defect movement (Q), jump distance, polarization loss constant, domain wall mobility and activation energy for domain wall movement. Curve fitting of the fatigue data on Pt/PZT/Pt capacitors was used to calculate the voltage and temperature dependence of the decay constant. The oxygen vacancy jump distance was calculated from the voltage dependence and found to be close to the lattice parameter value of the PZT films. The activation energy difference between domain wall movement and oxygen vacancy movement was calculated form the temperature difference of the decay constant and found to be 0.28222 eV for Pt/PZT/Pt capacitors at 4V [9-12].

The fatigue model discussed above provides some clear guidelines to minimize the degradation problem. If the relative movement of oxygen vacancies and their entrapment at the electrode-ferroelectric interface and/or at the domain boundaries and grain boundaries are the factors contributing to fatigue, then there are two possible solutions to overcome this problem : (1) reducing the tendency for defect entrapment by controlling the interface state and (2) controlling the defect density. These methods will be discussed separately in the forthcoming sections.

CONTROL OF INTERFACE STATE

The interface between a metal electrode such as Pt and an oxide ferroelectric such as PZT is chemically unstable as a result of the large lattice structure difference. In order to reduce the tendency for defect entrapment at the interface, the instability at the interface should be minimized. A plausible way to minimize fatigue then is to use an electrode material that has a lattice structure the same or similar to that of the bulk. This criterion is satisfied by conductive oxide electrodes such as RuO_2, indium tin oxide (ITO), lanthanum strontium cobaltate (LSCO) and yttrium barium copper oxide (YBCO). The conductive oxides can consume the oxygen vacancies by increasing their

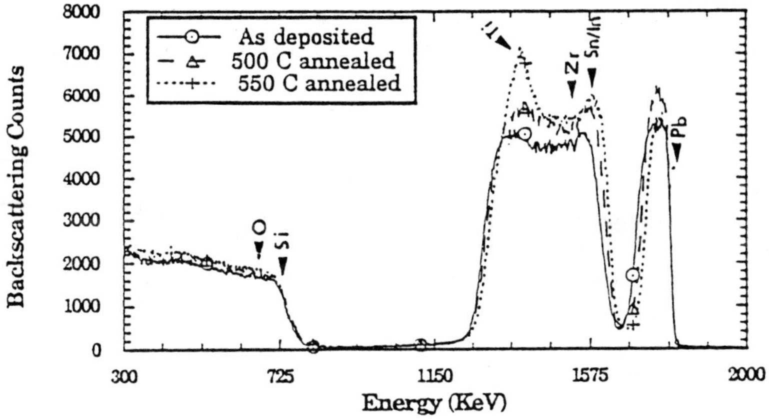

Figure 2(a) : RBS spectra of Si/ITO/PZT samples as a function of annealing temperature

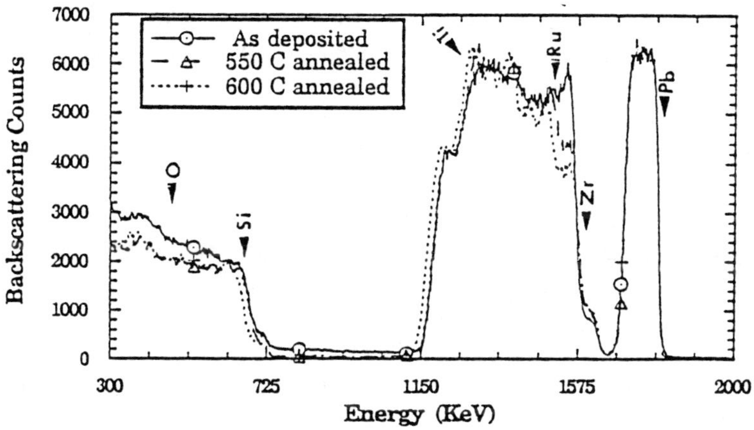

Figure 2(b) : RBS spectra of $Si/RuO_2/PZT$ samples as a function of annealing temperature

oxygen non-stoichiometry and thus prevent the development of space charge at the interface [13,17]. The metallic oxides of transition metals such as RuO_2, IrO_2, OsO_2 etc. tend to crystallize in the tetragonal rutile structure while LSCO and YBCO are perovskite metal oxides. These conducting oxides are promising candidates for various metallization schemes primarily because of their low resistivity (30-150 $\mu\Omega$-cm) and high thermal stability [18]. However, there are several other requirements that an electrode should meet for ferroelectric thin film applications.

The ferroelectric films are deposited on the bottom electrodes either at high temperatures or they are deposited at low temperatures and later annealed to form the desired ferroelectric phase (e.g., perovskite structure in PZT). At high temperatures interdiffusion of various elements of the electrode, ferroelectric and the underlying Si substrate is a major concern and may result in degradation of ferroelectric properties. Therefore, it is desirable to have an electrode that is a good diffusion barrier at high temperatures. Additionally, it is necessary to use an electrode that can be etched using the standard dry or wet chemical etching techniques that are being used in the VLSI industry today. For e.g., etching of Pt is a major concern if it used as an electrode for ferroelectric films.

We have investigated the effectiveness of two candidate electrodes - RuO_2 and ITO - for PZT thin film applications [17]. The ferroelectric properties and degradation of PZT films on ceramic and metal electrodes was thus compared. These electrodes were sputtered onto Si substrates and spin coated with sol-gel derived PZT thin films [17]. The samples were annealed at various temperatures between 400°C -700°C and characterized thereafter using the techniques of Rutherford backscattering spectrometry (RBS) and x-ray diffraction (XRD) to determine the extent of intermixing and the temperature of complete ferroelectric perovskite phase formation on the respective electrodes. The RBS profiles obtained from the Si/ITO/PZT and Si/RuO_2/PZT samples are shown as a function of annealing temperature in Fig. 2(a) and Fig. 2(b), respectively. In the Si/ITO/PZT samples, a considerable decrease in the Pb yield was observed above 500 °C indicating a loss of PbO. The drastic decrease in the low energy edge of Pb peak indicates that Pb diffuses into the ITO layer. In and Sn diffuse towards the surface and react with PbO to form a Pb-In-Sn compound. In the case of Si/RuO_2/PZT samples, no noticeable intermixing between Pb and RuO_2 was observed up to a temperature of 600 °C. Additionally, no significant change was observed in the Si edge indicating that the diffusion of Si into RuO_2 is minimal up to this temperature. However, increased intermixing between Pb and RuO_2 was observed in the samples annealed above 600 °C. XRD analysis indicated that the PZT perovskite formation is initiated at a temperature around 550 °C and complete at a temperature of 600 °C on Si/RuO_2 substrates while in the Si/ITO samples the PZT perovskite phase

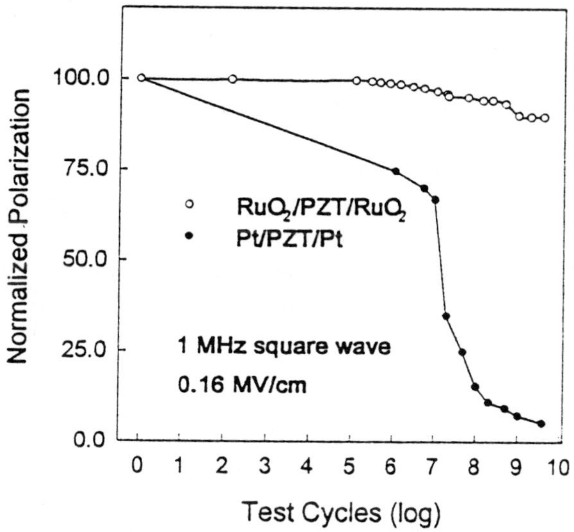

Figure 3(a) : Comparison of fatigue properties of Pt/PZT and RuO_2/PZT capacitors

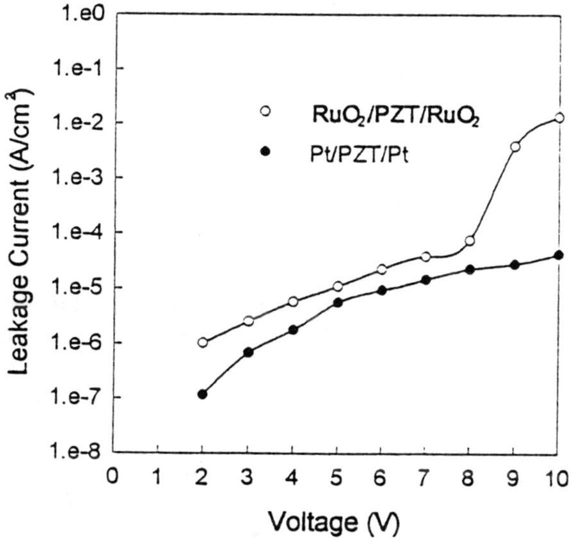

Figure 3(b) : Comparison of leakage current properties of Pt/PZT and RuO_2/PZT capacitors

was not observed even at a temperature of 700 °C. These results indicate that as far as diffusion barrier properties and perovskite phase formation temperature is concerned, RuO_2 is a better choice as an electrode for PZT thin films applications. The perovskite phase in PZT forms before any serious interdiffusion begins and therefore we are not limited by this factor in the processing of the device.

The hysteresis properties of RuO_2/PZT/RuO_2 capacitors, measured using the standardized RT66A ferroelectric tester, was in close comparison with that of Pt/PZT/Pt capacitors [17]. Typically, the hysteresis loops were well saturated at 5V and the films showed a P_r value of 20 $\mu C/cm^2$ and E_c value of 30 kV/cm. Fig. 3(a) compares the fatigue properties of RuO_2 and Pt electroded PZT films. While the Pt electroded films showed early loss of polarization, no fatigue was observed up to 10^{10} cycles with RuO_2 electrodes. Additionally, no change was observed in the dielectric permitivity measurements made on RuO_2/PZT/RuO_2 as a function thickness and number of fatigue cycles indicating the absence of any interfacial space charge layer build-up with increasing fatigue cycles [16]. Although RuO_2/PZT/RuO_2 capacitors did not fatigue, they showed higher leakage currents when compared to the Pt counterparts at similar electrical fields (Fig. 3(b)). One of the possible ways to minimize the leakage current in RuO_2/PZT capacitors is to use multilayer Pt/RuO_2 electrode structures [19]. It has been noticed that thin films (~ 20-70 nm) of RuO_2 grown on top of Si/SiO_2/Ti/Pt (111) substrates also provides a structural template layer for oriented growth of PZT films. Leakage currents in PZT films grown on Pt/RuO_2 electrodes is lower by at least an order of magnitude compared to RuO_2/PZT capacitors [19]. Other candidate oxide electrode materials such as LSCO and YBCO are also known to reduce the fatigue rate in PZT films [20, 21].

CONTROL OF DEFECT DENSITY

In an earlier section, it was shown that point defects such as oxygen vacancies can be created during the processing of oxide ferroelectric thin films (e.g, PZT) at high temperatures. Considering the case of PZT, since the vacancy concentrations resulting from PbO volatilization can be very high, techniques for controlling the PbO activities during film formation are very important for minimizing defect concentrations. One common technique is to provide excess PbO during film formation to compensate the loss. However, precise control is often very difficult and chemical segregation problems at various structural interface sites such as grain boundaries are commonly observed. In addition to compensating the PbO loss, another technique that can be used to lower the oxygen vacancy concentration is the introduction of donor dopants (e.g., Nb^{5+} at Zr or Ti sites). The consumption of oxygen vacancies can then be written as [13]:

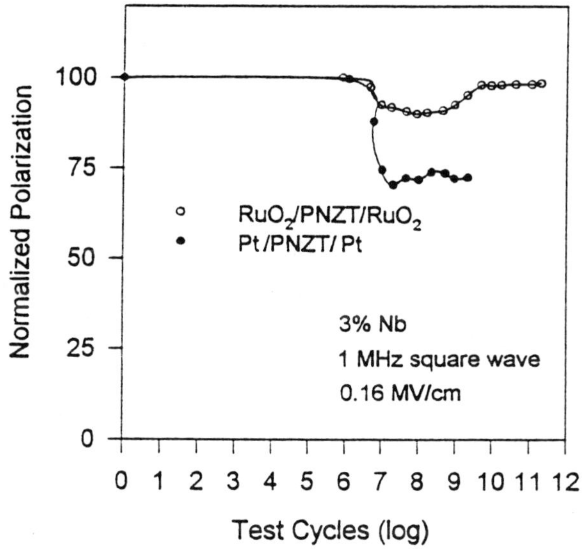

Figure 4(a) : Fatigue properties of Nb-doped Pt/PZT and RuO_2/PZT capacitors

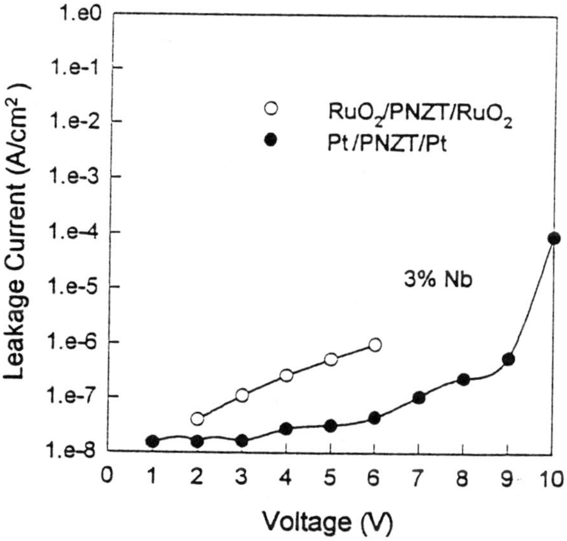

Figure 4(b) : Leakage current properties of Nb-doped Pt/PZT and RuO_2/PZT capacitors

$$Nb_2O_5 + V_O^{oo} \rightarrow 2Nb_{Ti,Zr}^{o} + 5O_O \qquad \text{------------ (8)}$$

where $Nb_{Ti,Zr}^{o}$ represents a Nb^{5+} ion at $(Ti, Zr)^{4+}$ site in PZT. In addition to reducing the oxygen vacancies, the donor dopants can also compensate for any kind of acceptors present in the films as shown below :

$$A_2O + Nb_2O_5 \rightarrow 2A_{Pb}' + 2Nb_{Ti}^{o} + 6O_O \qquad \text{------------ (9)}$$

Using this approach, Nb and La-doped PZT (PNZT and PLZT) films were fabricated by sol-gel/spin coating on RuO_2 and Pt electrodes. Fig. 4(a) depicts the fatigue behavior of Nb doped PZT films on both RuO_2 and Pt electrodes. By comparison (with Fig. 3(a)), it is clear that the fatigue behavior of Pt electroded PNZT films is much better than that of Pt electroded PZT films. Furthermore, as shown in Fig. 4(b), the leakage currents of PNZT films are lower than those of PZT films for both Pt and RuO_2 electrodes (compare with Fig. 3(b)).

The minimization of intrinsic defect concentration may also be accomplished by choosing compounds with inherently high defect formation energies or compounds that have no volatile components in their sublattice exhibiting polarization. In the case of perovskite type ferroelectrics such as PZT, the entire lattice is responsible for the ferroelectric phenomena; the volatile PbO component is a part of the perovskite lattice contributing to the polarization. Identification of alternate ferroelectric materials (including structure types other than the perovskite) that do not contain any volatile species in its sublattices exhibiting polarization is therefore another solution to overcome the problem of fatigue. It is to be noted that for nonvolatile random access memory applications, the chosen ferroelectric materials should also satisfy certain basic criteria such as room temperature stability, high Curie temperature, large P_r and P_s, low E_c, low dielectric constant, low dielectric loss and finally, easy thin film processibility using standard VLSI processing techniques.

In addition to perovskite type ferroelectric oxides, other structure types that exhibit ferroelectricity include : pyrochlore, tungsten bronze and layer-type bismuth compounds [22-24]. Using the criteria stated above and noting the presence/absence of any volatile component in the ferroelectric sublattice, all but a few oxides belonging to the $Sr_2Nb_2O_7$ (pyrochlore) family, $YMnO_3$ (perovskite) family and the layer-structure family were eliminated. Thin films of these compounds were grown in situ by pulsed excimer laser ablation on platinized Si/SiO_2 substrates. This technique has become particularly attractive ever since it was used to successfully deposit *in situ* multicomponent high temperature superconducting oxide thin films [25]. It offers the advantages of multicomponent composition and phase control in complex oxides, *in-situ* epitaxial growth of thin films, compatibility with semiconductor technology and ability to deposit a wide variety of materials over a broad pressure regime [26].

Although $Sr_2Nb_2O_7$ and $La_2Ti_2O_7$ are promising candidates for this application as a result of their high Curie temperature, large spontaneous polarization and low coercive field values in the bulk form, low temperature processing of these oxides in the ferroelectric phase was found to be difficult. The ferroelectric properties of $YMnO_3$ thin films could not be determined in the thin film form as a result of the large leakage current exhibited by these films in the range of the applied voltages (1-5 V). The high leakage current is possibly due to the multiple oxidation states of manganese. In the layer-structure family, a large number of compounds of the general form $(Bi_2O_2)^{2+}$ $(M_{n-1}R_nO_{3n+1})^{2-}$, where M= Ba, Pb, Sr, Bi, K, or Na, n= 2,4, or 5 and R = Ti, Nb or Ta, are known to be ferroelectric [23-24]. These compounds have a pseudo-tetragonal symmetry and the structure is comprised of stacking of n perovskite-like units of nominal composition MRO_3 between Bi_2O_2 layers along the pseudo-tetragonal c-axis [23-24]. For our investigation, Pb-based (e.g., $PbBi_2Nb_2O_9$, $PbBi_2Ta_2O_9$) compounds were not considered because Pb atoms are a part of the perovskite sublattice responsible for spontaneous polarization in these materials. PbO volatility at high temperatures is therefore still a problem. The K and Na based oxides were eliminated because of their poor room temperature stability. Furthermore, compounds that contain Bi^{3+} ions in the perovskite-like units were also not considered because Bi_2O_3 could volatilize during high temperature processing and undesirable oxygen vacancies could be created in the perovskite lattice leading to fatigue. For example, $Bi_4Ti_3O_{12}$ is a well characterized layer-structure ferroelectric oxide in both the bulk and thin film form. However, thin films of $Bi_4Ti_3O_{12}$ need to be processed at high temperatures (> 700 °C) to obtain the ferroelectric phase. At these temperatures, Bi_2O_3 (part of the perovskite-like units in this compound) is volatile, resulting in creation of oxygen vacancies in the material. This compound is known to show fatigue on Pt electrodes [27]. The investigation of possible fatigue free ferroelectrics was thus restricted to compounds that contain Ba or Sr in the M site with Ta or Nb in the R site and solid solutions of these compounds. The Sr based compounds have higher Curie temperatures compared to Ba based compounds and therefore show the stronger domain interactions that are needed to overcome other degradation problems such as retention. Thus, greater emphasis was placed on the investigation of compounds belonging to the $SrBi_2Nb_2O_9$ - $SrBi_2Ta_2O_9$ (SBTN) solid solutions for possible fatigue free applications.

$SrBi_2(Ta_xNb_{1-x})_2O_9$ ($0<x<1$) films (SBTN) of various compositions were deposited onto $Si/SiO_2/Ti/Pt$ substrates by PLD (KrF, 248 nm) under the conditions : substrate temperature = 700 °C, O_2 partial pressure = 200 mTorr, pulse energy = 700 mJ/pulse and pulse frequency =10 Hz. The as-deposited films were specular and the measured composition of the films (by electron microprobe x-ray analysis) were close to the desired stoichiometry. X-ray diffraction studies indicated the predominant orientation of these films to be along the (115) and (00*l*) planes as shown in Fig.5. The hysteresis properties of the films with were measured with Pt top electrodes using the standardized RT66A ferroelectric tester (Radiant Technologies). The top electrodes were deposited by RF sputtering using a shadow mask to an area of 2.1×10^{-4} cm^2.

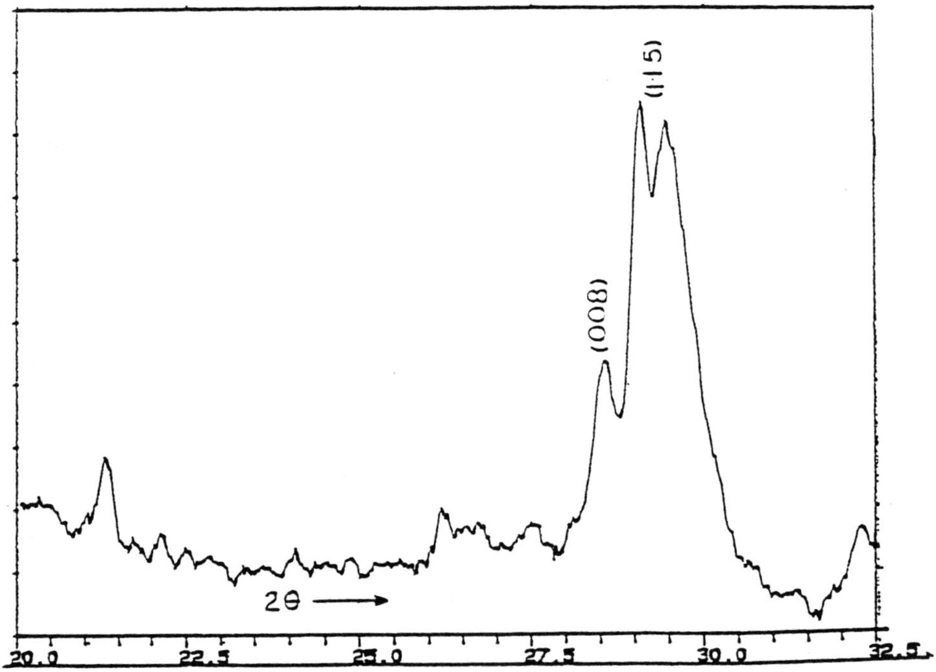

Figure 5 : XRD pattern of $SrBi_2(Ta_{0.4}\ Nb_{0.6})_2O_9$ films grown on $Si/SiO_2/Ti/Pt$ at 700 °C

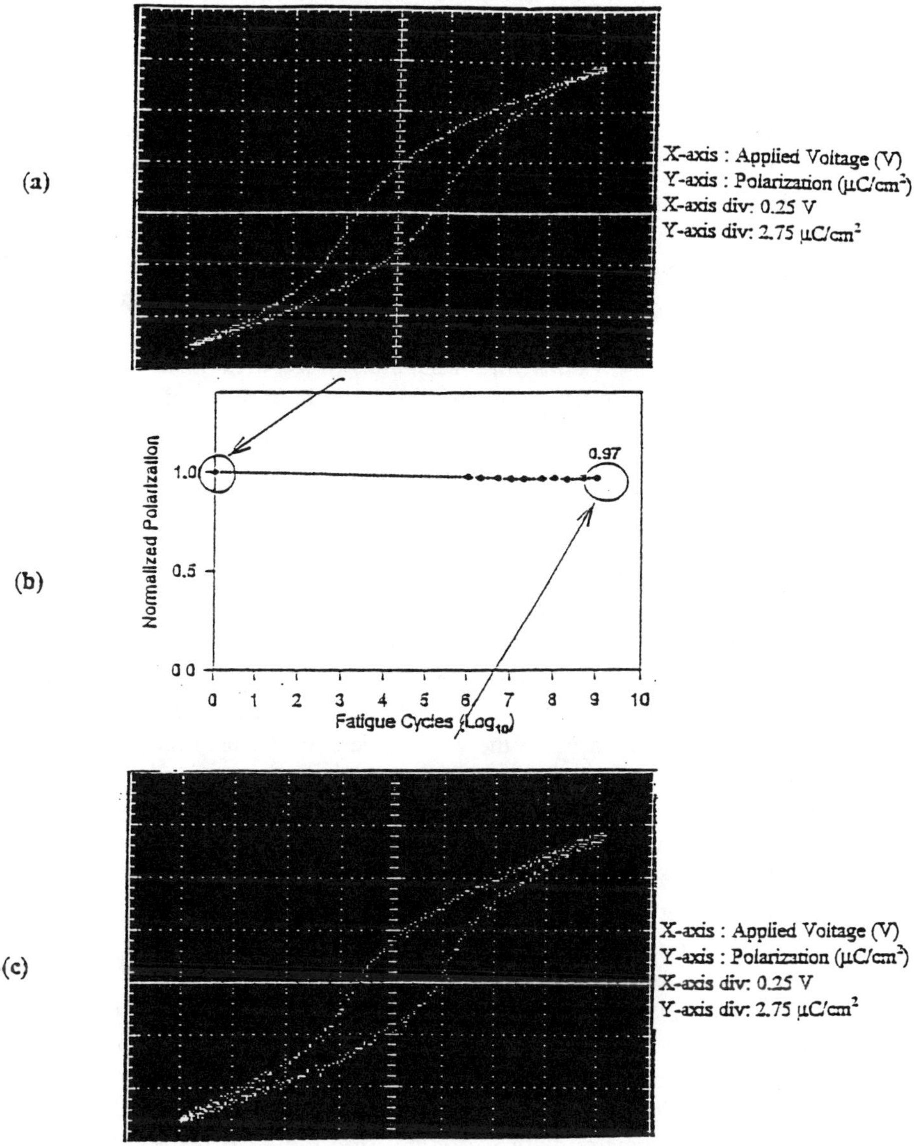

Figure 6 : Hysteresis and fatigue properties of $SrBi_2(Ta_{0.4}\,Nb_{0.6})_2O_9$ films grown on $Si/SiO_2/Ti/Pt$ at 700 °C

Figure 6(a) shows the hysteresis characteristics of as-deposited SBTN (x=0.4) films at an applied voltage of 5V. The hysteresis loop was well saturated at this voltage; a $2P_r$ value of 22 $\mu C/cm^2$ and an E_c value close to 60 kV/cm were obtained. These values are well within the requirements for nonvolatile memory device applications. The fatigue test on the films were performed using a 5V square wave a.c. signal at a frequency of 1 MHz input through a pulse generator. Figure 6(b) shows the results of the fatigue study performed up to 10^{10} cycles. The films did not show any significant fatigue up to this cycling; there was only a 3 % loss in polarization after 10^{10} cycles. This was confirmed by measuring the hysteresis properties of the films after fatigue as shown in Figure 6(c); the hysteresis properties of the films before and after fatigue cycling were similar. The leakage current density values for these films were measured as a function of applied voltage. At an applied voltage of 3V, the films showed a leakage current density value less than 10^{-7} A/cm^2. The measured resistivity of films were typically of the order of 5×10^{12} Ω-cm [28].

The polarization in layered structure compounds are dependent on the orientation. Additionally, the measured coercive field and the resistivity of the films are also a function of the orientation. A well known example that illustrates this point is the case of $Bi_3Ti_4O_{12}$ [27]. $Bi_3Ti_4O_{12}$ thin films (layered structure) exhibit larger spontaneous values when oriented along the a-axis compared to orientation along the c-axis. However, the coercive field values along the a-axis are also very large compared to the c-axis orientation which makes the switching process in these materials difficult. In our case, since we obtained strong (115) and (00*l*) orientations in the films, the resultant polarization was a combination of the c and a-axis polarizations. Based on the observations in $Bi_3Ti_4O_{12}$, it was reasonable to expect SBTN films deposited with a preferred c-axis orientation would exhibit lower coercive fields and higher resistivities albeit at the expense of lower polarization values.

For favorable growth of SBTN films along the c-axis, they were deposited on MgO (100)/Pt(100) substrates. Pt(100) has a close lattice match (0.389 nm) to the 'a' lattice parameter of the SBTN unit cells. Fig. 7 shows the typical XRD scan for SBTN films with a composition close to x=0.4. A number of predominant diffraction peaks associated with scattering from the (00*l*) planes indicates clearly that the preferred orientation in this case is along the c-axis. Fig. 8 shows the hysteresis properties of SBTN films (x=0.4) at an applied voltage of 5 V. The films showed a remnant polarization of 11 $\mu C/cm^2$ and a coercive field of 45 kV/cm. Fatigue tests conducted using a 5 V bipolar square wave at a frequency of

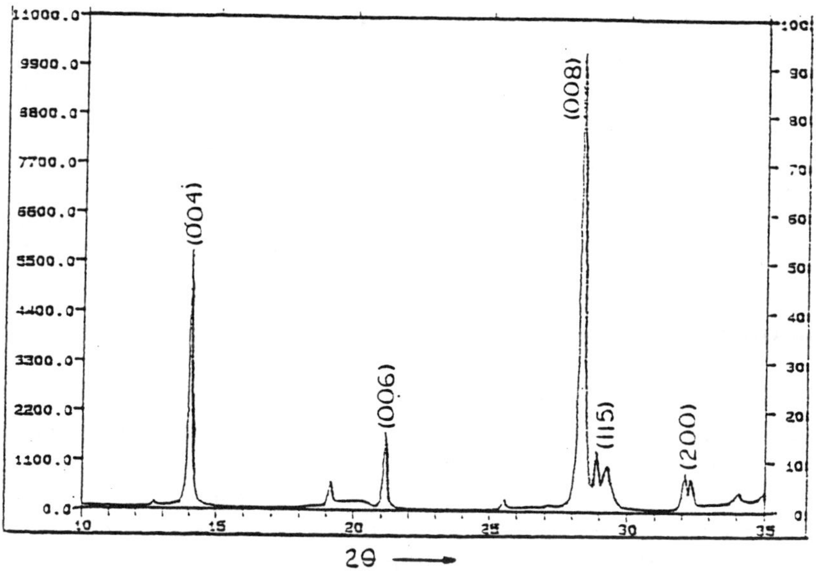

Figure 7 : XRD pattern of $SrBi_2(Ta_{0.4}\,Nb_{0.6})_2O_9$ films grown on MgO (100)/Pt (100) substrate at 700 °C (c-axis oriented films)

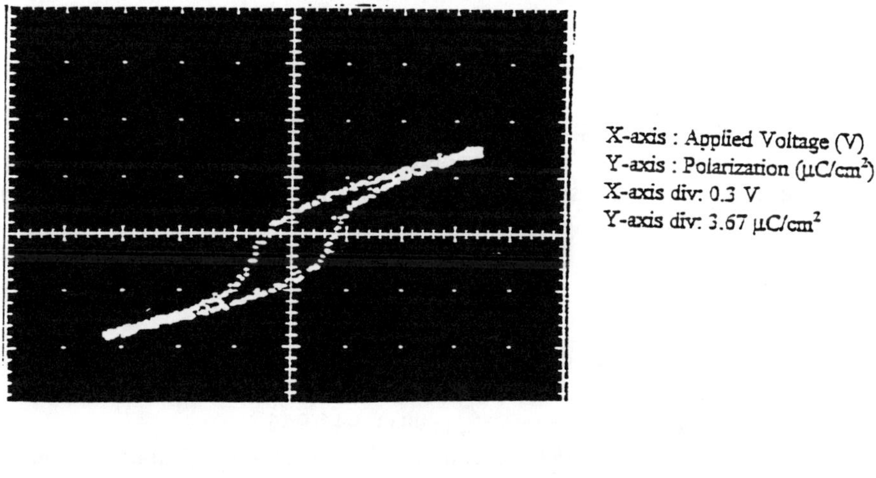

Figure 8 : Hysteresis properties $SrBi_2(Ta_{0.4}\,Nb_{0.6})_2O_9$ films grown on MgO(100)/Pt (100) substrate at 700 °C (c-axis oriented films)

1 MHz did not indicate any significant loss of polarization up to 10^{10} test cycles. Additionally, the films showed a very low resistivity value of 4×10^{13} Ω-cm. Comparison of these properties to the earlier results obtained on films with no preferred c-axis orientation indicate similar remnant polarization values but significantly lower coercive fields (a decrease by 15 kV/cm) and an increase in resistivity by at least an order of magnitude. The improved properties clearly indicate that is preferable to grow c-axis oriented films for nonvolatile memory applications.

SUMMARY

Fatigue in ferroelectric films has been attributed to domain pinning by space charge at the electrode-ferroelectric interfaces. The role of oxygen vacancies in the development of the space charge at the interfaces has been discussed. Based on a quantitative fatigue mechanism, it was shown that fatigue can be minimized by : (1) controlling the interface state by using conductive oxide electrodes such as RuO_2 (for PZT films) which can consume the oxygen vacancies at the electrode and ferroelectric interface and thus eliminate the space charge development and (2) controlling the defect (mainly oxygen vacancies) density using donor dopants such as Nb in PZT that reduces the oxygen vacancy concentration or by utilizing ferroelectric films such as $SrBi_2(Ta_xNb_{1-x})_2O_9$ which have inherently low defect concentration.

ACKNOWLEDGMENTS

This project was partially supported by ARPA through ONR, CERAM Inc.,VA and Sharp Co., IC group. Additional funding was also obtained from CIT, VA. The author would also like to thank Dr. I.K. Yoo (Samsung, Korea) and members of the Thin Films group, Virginia Tech.

REFERENCES

1. J. F. Scott and C.A. Paz de Araujo, Science, 246, (1989), 1400-1405.
2. R. Ramesh, A. Inam, W.K. Chan, B. Wilkens, K. Myers, K. Remschang, D.L. Hart, J.M. Tarascon, Science, 252, (1991), 944-946.
3. L. Parker and A. Tasch, IEEE Circuits and Devices, pp.17, January 1990.
4. Anonymous, Microprocessors and Microsystems, 13, 291, (1989).
5. C. Karan, *IBM Tech Report*, 1955.
6. I.S. Zheludev, Physics of Crystalline Dielectrics, Electrical Properties, Plenum,

New York, Vol.2, 1971, pp. 474-490.
7. A.Y. Kudzin, T.U. Panchenko and S.P. Yudin, Sov.Phys. Solid State, 16(8), (1975), 1589.
8. H.M. Duiker, P.D. Beale, J.F. Scott, C.A. Paz de Araujo, B.M. Melnick, J.D. Cuchlaro and L.D. McMillan, J.Appl. Phys., 68(11), (1990), 5783.
9. I.K. Yoo and S.B. Desu, Mat. Sci. and Eng., B13, (1992), 319.
10. I.K. Yoo and S.B. Desu, Phys. Stat. Sol., a133, (1992), 565.
11. I.K. Yoo and S.B. Desu, J. Int. Mat. Sys., 4, (1993), 490.
12. S.B. Desu and I.K. Yoo, J. Electrochem. Soc., 140, (1993), L133.
13. S.B. Desu, D.P. Vijay and I.K. Yoo, Mat. Res. Soc. Symp., 335, 53, (1994).
14. J.F. Scott, C.A. Araujo, B.M. Melnick, L.D. McMillan, R. Zuleeg, J. Appl. Phys. 70 (1), (1991), 382.
15. C.K. Kwok and S.B. Desu, Mat. Res. Soc. Symp. Proc., 243, 393, (1992).
16. J.J. Lee and S.B. Desu, Mat. Res. Soc. Symp. Proc., *Fall Meeting,* (1994).
17. D.P. Vijay and S.B. Desu, J. Electrochem. Soc, 140, (1993), 2640.
18. L. Krusin-Elbaum and M. Wittmer, J. Electrochem Soc., 135(10), 2610, (1988).
19. I.L. Chung, S.B. Desu and D.P. Vijay, Mat. Res. Soc. Symp. Proc., *Fall Meeting,* (1994).
20. R. Ramesh, J. Lee, T. Sands, and V.G. Keramidas, Appl. Phys. Lett. 64 (19), 2511, (1994).
21. R. Dat, J. Lichtenwalner, O. Auciello, and A.I. Kingon, Appl. Phys. Lett. 64 (20), 2673, (1994).
22. B. Aurivillius, Arkiv Kemi, 1[54], (1949), 463.
23. E.C. SubbaRao, J. Phys. Chem. Solids, 23, (1962), 665.
24. G.A. Smolenski, V.A. Isupov and A.I. Agranovskaya, Fiz Tverdogo Tela, 3[3], (1961), 895.
25. D. Dijkkamp, T. Venkatesan, X.D. Wu, S.A. Saheen, N. Jisrawi, Y.H. Min-Lee, W.L. Mclean and M. Croft, Appl. Phys. Lett., 51, (1987), 619.
26. P.C. Joshi and S.B. Krupanidhi, J. Appl. Phys., 72(12), (1992), 5827.
27. N. Maffei and S.B. Krupanidhi, J. Appl. Phys, 72(8), (1992), 3617.
28. S.B. Desu and D.P. Vijay, Mat.Sci.and Engg., *accepted for publication*, (1994).

DESIGN AND DEVELOPMENTOF A LOW K SILICA DIELECTRIC

Tapan K Gupta
Westinghouse Electric Corporation
Electronic Systems Group
Baltimore, Maryland 21203

and

Jau-Ho Jean
National Tsing Hua University
Hsinchu
Taiwan, ROC

ABSTRACT

A silica dielectric containing a mixture of a high-softening high silica glass (HSG) and a low-softening borosilicate glass (BSG) delivers a dielectric constant of 4.0 - 4.2, but its fabrication is hindered by the formation of cristobalite that undergoes phase transformation during cooling. It is shown that the key to preventing cristobalite formation is to add a small amount of alumina to the mixture of glasses. When sintered, this mixture produces a phase transformation-free, single phase silica dielectric that is amorphous to XRD. The shrinkage kinetics conform to a model of liquid phase sintering wherein the rate controlling step is the viscous flow of BSG. This low k silica dielectric in combination with gold metallurgy presents an attractive array of properties for multilayer packaging application.

INTRODUCTION

The driving forces for the advanced ceramic electronic package are higher characteristic speed of the media, smaller feature size of the conductors, greater reliability against mechanical, thermal and environmental perturbations and continuing lower cost. The way of enhancing the speed through interconnections between VLSI chips lies in improving the media speed and reducing the wiring

To the extent authorized under the laws of the United States of America, all copyright interests in this publication are the property of The American Ceramic Society. Any duplication, reproduction, or republication of this publication or any part thereof, without the express written consent of The American Ceramic Society or fee paid to the Copyright Clearance Center, is prohibited.

length. In this paper, we choose to concentrate only on the characteristic that iimproves the media speed. The material property that controls the media speed (v) is the dielectric constant (k) given by the relation: $v = c/\sqrt{k}$, where c is the speed of the light. Thus, by decreasing the dielectric constant of the substrate, the signal propagation through interconnections can be increased. Of particular concern to the designer of advanced ceramic packages is that the k of alumina (Al_2O_3), a widely used packaging material, is about 9, hence the speed of propagation of a signal in alumina is only 1/3 that of a perfect material with k = 1 (vacuum). Reducing the dielectric constant to about 4, that of silica (SiO_2), represents an increase in speed of about 50%, from 1/3 to 1/2 of that in vacuum. However , there exists a major problem with SiO_2 that has prevented its widespread use as a pure dense monolithic structure in the package. It comes in three crystalline varieties: quartz, tridymite and cristobalite, all of which undergo displacive phase transformation during cooling from high temperature, resulting in volume expansion and cracking [1]. As a result packages consisting mainly of SiO_2 have never been made into a practical product [2,3], although it remains a great attraction should the phase transformation be prevented during cooling. The use of an amorphous silica avoids the issue of crystallinity and phase transformation. In this investigation, we have, therefore, chosen a high-softening, high silica glass (HSG) as a substitute for crystalline ceramic, and used a borosilicate glass (BSG) to flux-sinter the mixture at a temperature of less than 1000 ^{0}C, so that noble metals and copper can be co-sintered with the dielectric. Since both these glasses have low dielectric constant of about 4 (see Table 1), the mixture rule [1] suggests that the resultant product will also exhibit a dielectric constant of about 4, that of silica.

Table 1. Properties of BSG and HSG

Glass	Density g/cm^3	k	TCE ppm/ ^{0}C
BSG	2.13	4.0	3.5
HSG	2.18	3.9	0.75

RAW MATERIALS

Both BSG and HSG glasses were obtained from commercial sources and used as received. Compositions of the glass powders were analyzed by atomic absorption (AA) and inductively coupled plasma (ICP). The results showed that the BSG contained approximately 75 wt% SiO_2, 25 wt% B_2O_3, and trace amount of alkali oxides, such as Na_2O and K_2O; and HSG, 95 wt% SiO_2 and 5 wt% B_2O_3. In addition, the BSG and HSG had softening points of 700 - 750, and >1500 ^{0}C, respectively. Particle size distributions of the powders, measured by sedimentation, showed a median size of about 6.5 μm. XRD analysis of the as-received glass powders showed amorphous patterns. The composition selected for the present study was obtained from previous study and consists of 60 vol% HSG and 40 vol% BSG [4].

DEVITRIFICATION OF BINARY GLASSES AND ITS PREVENTION

In developing a useful, phase transformation-free dielectric from a binary mixture of HSG and BSG, it is important to determine whether the mixture will yield a crystalline SiO_2 on sintering. This concern arises because evidence exists [5] that fused silica devitrifies to cristobalite when held at a high temperature for a period of time. To resolve this concern, pellets were made from a composition of 60 vol% HSG and 40 vol% BSG with a diameter of 1.9 cm and a height of 0.3 cm, by pressing the powder at 9 x 10^6 N/m^2 and sintering at temperature between 825 to 950 ^{0}C. No crystallinity was observed when sintered for a short time, from 15 to 30 minutes, but when the samples were held for a longer period of time, cristobalite was indeed detected in the sintered body by XRD. Figure 1 shows the X-ray intensity data of a composition containing 60 vol% HSG and 40 vol% BSG and sintered at 925 ^{0}C for various periods of time. It was noted elsewhere [6] that the crystallization takes place only in the binary mixture of HSG and BSG, with end members showing no devitrification when held at temperatures between 800 and 1000 ^{0}C for 4 hours. It was further noted that the precipitation of cristobalite is characterized by an incubation period and that it originates in HSG particle.

However, the rate limiting step is determined to be the transport of alkali ions in BSG.

Given that the above reaction kinetics prevail in the binary glass mixture, the strategy for preventing the cristobalite formation was to divert the transport the alkali ions from the desired target to a more attractive but harmless target and to do so in a time frame shorter than the characteristic incubation period of the cristobalite growth. It was determined [7] that addition of alumina served as a crystal growth inhibitor to a mixture of HSG and BSG under the conditions prevailed during sintering. The inhibition occurred [8] as a result of forming a reaction layer around each alumina particle, caused by a strong coupling between Al^{3+} from Al_2O_3 and alkali ions (K^+, Na^+) from BSG, thus denying the HSG its vital rate limiting step of K transport that is necessary to form cristobalite.
The effects of alumina on the devitrification of a binary mixture of HSG and BSG are shown in Fig 2 where 1 to 10 vol% of Al_2O_3 is added to the composition of 60 vol% of HSG and 40 vol% of BSG by replacing an equivalent amount of HSG. Disks samples were prepared as before and sintered at temperature between 800 and 1400 ^{0}C for 8 hours. Note that cristobalite appears only in the absence of alumina in the temperature range of 800 to 1000 ^{0}C, whereas cristobalite-free phase zones exist with various amounts of Al_2O_3. The description of these phases are described elsewhere [7].

SINTERING OF A BINARY MIXTURE OF HSG AND BSG

Having found a way to prevent cristobalite formation in an otherwise vulnerable mixture of HSG and BSG, investigation was carried out to determine the sintering kinetics of a mixture of 60 vol% HSG and 40 vol% BSG with a small addition of alumina which was substituted for HSG. Samples used in this study was prepared by tape casting and lamination process. The organic composition for tape casting contained binder, plasticizer, dispersant and solvents. The total binder content was approximately 25 wt% of the inorganic powders. The mixture of organic and inorganic compounds was ground for 16 h in a ball mill to a median particle size of 2 - 4 μm, de-aired and tape-caste on a cellulose acetate carrier. The tapes were uniform and had a thickness of about 200 μm. Samples with an approximate thickness of about 2000 μm were made by placing 10 - 11 green sheets between two metal plates and applying pressure from top and bottom using a laminator. The approximate pressure was 4.2 - 21 x 10^6 N/m^2. workable temperature in the range of 60 -100 ^{0}C, , and time 30 - 300 s. The relative green density of pressed laminates was measured dimensionally and ranged from 55 to 57%. The laminates were then sectioned to a size of 3 x 2 x 15 mm, and the linear shrinkage during sintering was continuously monitored [4] by a thermomechanical analyzer (TMA). A nominal load of 0.01N was applied to maintain an intimate contact between probe and sample during firing. The organic binder was removed by taking the sample from room temperature to 500 ^{0}C in flowing air at 3 ^{0}C/min, and holding at 500 ^{0}C for 60 min. A small dimensional shrinkage, 1 - 1.3%, was detected during this binder burnout. Taking this dimensional change as the base line, isothermal sintering was conducted from 500 ^{0}C to various sintering temperatures ranging from 875 to 950 ^{0}C, using a heating rate of 50 ^{0}C/min in air.

A typical plot showing the results of fractional shrinkage, $\Delta L/L_0$, as a function of time (t) at various temperatures, 875, 900, 925, 950 ^{0}C is illustrated in Fig. 3. Clearly, the sintering behavior follows a similar pattern for all temperatures investegated: the shrinkage increases rapidly initially and then slows down to a nearly constant shrinkage of about 16%. When these data are replotted as $\ln(\Delta L/L_0)$ vs ln(t), straight lines are obtained up to a shrinkage of 14% at all temperature investigated as shown in Fig. 4, indicating that the shrinkage behavior of this system can be described by the liquid phase sintering kinetics as proposed by Kingery [9]:

$$(\Delta L/L_0)^n = Kt/T \qquad (1)$$

where n is an exponent, K the temperature depedent rate constant and T the absolute temperature. The Eqn. (1), on rearrangement, yields:

$$n \ln(\Delta L/L_0) = \ln(K/T) + \ln(t) \qquad (2)$$

which agrees with the plot of Fig. 4.

Using the least-square fit method, the values of n, calculated from the slopes (1/n) of Fig. 4 are 2.84 at 875, 3.34 at 900, 3.20 at 925 and 3.23 at 950 ^{0}C. Assuming that these slight variation in n values does not signify a change in sintering mechanism, a mean value of 3.15 is used in Eq. (1) to calculate the activation energy, by using the logarithms of the specific rates T/t versus 1/T. As shown in Fig. 5, a family of straight lines with similar slopes are found at various linear shrinkages from 5 to 14%. The apparent activation energies, obtained from the slopes of the straight lines, lie between 175 - 230 KJ/mole, which is in reasonable agreement [4] with the activation energy range for the viscous flow of BSG, 220 - 240 KJ/mol. From this we conclude that the sintering is controlled by the viscous flow of BSG while HSG acts as a high temperature insulator.

PROPERTIES OF THE NEW DIELECTRIC

The new dielectric is an amorphous material with no trace of crystallinity as determined by XRD. This dielectric exhibits an excellent properties for conventional and microwave applications. Off-chip delay, obtained from microstrip measurement using a gold conductor, is 55 ps/cm. The most important electrical characteristics, the dielectric constant and tan δ loss as a function of frequency, are shown in Fig. 6 while the physical and electrical data are presented in Table 2. Since the new dielectric is formed by liquid phase sintering, it conforms to the shape of the setter plate and with a flat setter plate, yields a very low camber. It can be co-fired with internal and external gold metallurgy, and is compatible with internal Ag/Pd and Ag. No experiments are conducted with Cu metallurgy in reducing atmosphere, so results are unknown. Low camber and shrinkage matching with metal make tight tolerance design possible. Multilayer packages have been built with or without cavity. External metal layer and resistor element can be be post-fired on to the surface of the substrate. Since it has a dielectric constant of about 4.0, it provides a challenge to the high end of polyimide dielectric with a dielectric constant of 3.5-4.0.

Table 2. Physical and Electrical Properties of Low k/Au System

Parameters	Low k SiO_2/Au
Dielectric	
Dielectric constant (KHz)	3.9 - 4.2
Dissipation factor (KHz)	0.1 %
Volume resistivity	$>10^{16}$ Ω-cm
Expansion coefficient (Si = 3 ppm/ °C)	2 - 6 ppm/ °C
Thermal conductivity	1.3 W/m-°C
Moisture absorption	< 0.1%
Density	2.2 g/cm^3
Color	White
Strength	12 - 15 ksi
Weibull exponent	> 10
Insulation resistance	10^{12}
Dielectric strength	34kV/mm
Decomposition/melting temperature	1710 °C
Package	
Conductor resistivity	2.2 ohm-cm
Line width	4 - 6 Mil*
Line thickness	10 Micron
Printing technology	Screen printing
Sheet resistance	1.6 mΩ/sq
Off-chip delay (Microstrip measurement)	55 ps/cm
Camber	≤ 1 mil/in
C0-firing temperature	< 1000 °C
Co-firing atmosphere	Air
Package Support	Stand alone package
Co-firing on Si Substrate	yes
TCE match to Si substrate	Excellent
Flip-chip die attach	No stress
Plating	May not be required

*Process limited at present, design optimum 25 Microns

It has a lower tan δ loss, and a 4-5 times higher thermal conductivity than that of polyimide. It has also a higher mechanical strength than that of polyimide and is of the order of 12-15 ksi with a Weibull exponent of greater than 10. It adheres to Si, so that multilayer packages can be built on Si substrate which can dissipate heat by vittue of its high thermal conductivity.

ACKNOWLEDGMENT

The authors acknowledge the assistance of Bob Sutterlin. Bill Straub, Guo-Quan Lu and Bill Moffatt in developing this dielectric.

REFERENCES

1. W. D. Kingery, H. K. Bowen and D. R. Uhlmann, Chap. 12 - 15 and 18 in *Introduction to Ceramics*, John Wiley & Sons, New York, 1976.
2. R. R. Tummala, " Ceramic Packaging", Chap. 7 in *Microelectronic Packaging Handbook*, Ed. by R. R. Tummala and E. J. Rymaszewski, Van Nostrand Reinhold, New York, 1989.
3. J.-H. Jean and T. K. Gupta, "Liquid-phase Sintering of Glass-cordierite System: Particle Size Effect," J. Mat. Sci., 27 (1992) 4967-4973.
4. J.-H. Jean and T. K. Gupta, "Isothermal and Nonisothermal Sintering Kinetics of Glass Filled Ceramics," ibid., 7 (12) 3342-3347 (1992).
5. N. G. Ainslie, C. R. Morelock and D. Turnbull, "Devitrification Kinetics of Fused Silica" Symposium on Nucleation and Crystallization in Glass and Melts, Ed. by M. R. Reser, G. Smith and H. Insley, Am Cerm. Soc. 97-107, 1962.
6. J.-H. Jean and T. K. Gupta, "Crystallization Kinetics of Binary Borosilicate Glass Composite," J. Mat. Res. 7 (11) 3103-3111 (1992).
7. J.-H. Jean and T. K. Gupta, "Devitrification Inhibitor in Binary Borosilicate Glass Composote," ibid., 8 (2) 356-363 (1993).
8. J.-H. Jean and T. K. Gupta, "Kinetics of Interfacial Reaction Between Borosilicate Glass and Sapphire Substrate," J. Mat. Res. 7 (9) 2514-2520 (1992).
9. W. D. Kingery, J. Appl. Phys. 30, 301 (1959)

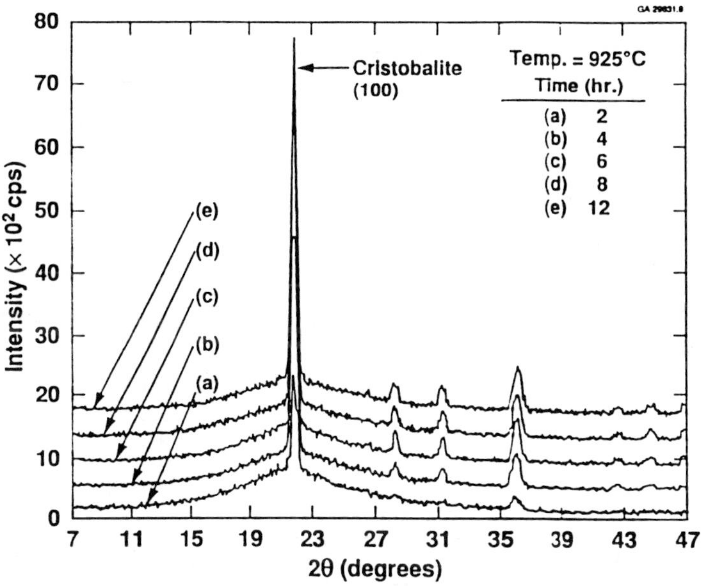

Figure 1. XRD results of a mixture of 40 vol % BSG and 60 vol % HSG sintered at 925 °C for 2, 4, 6, 8 and 12 hours.

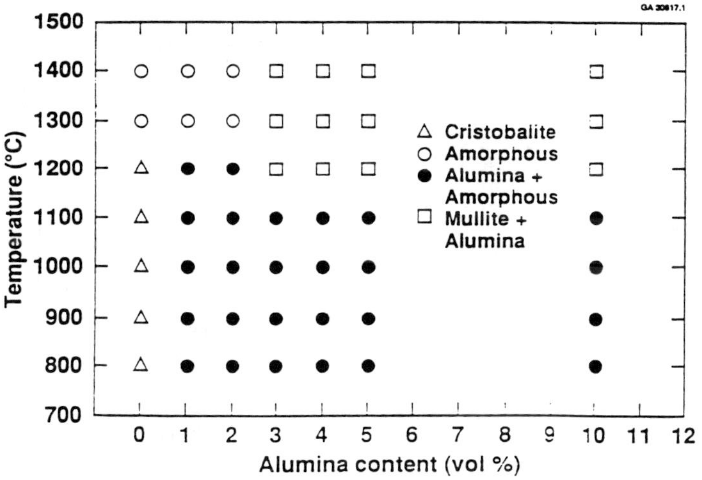

Figure 2. Summary of XRD results for the samples with various alumina contents sintered at temperatures from 800 to 1400 ^{0}C for 8 h.

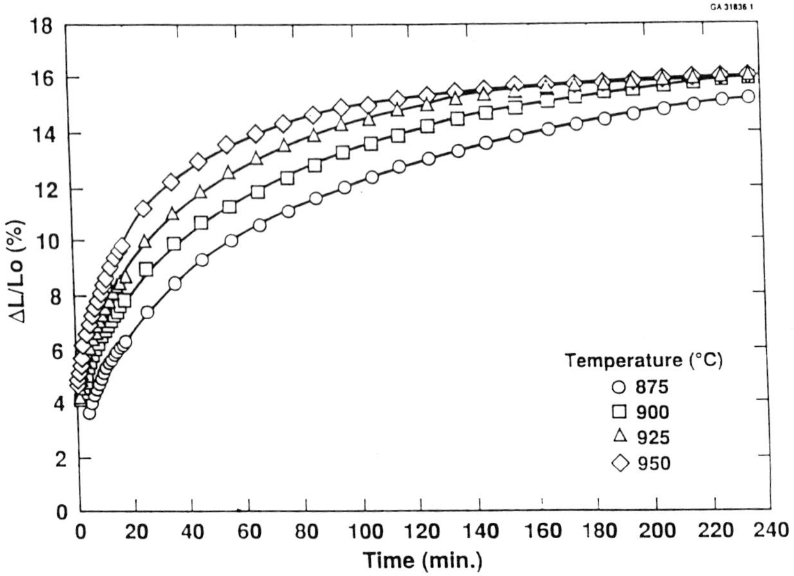

Figure 3. Linear shrinkage versus time obtained at a constant sintering temperature of 875, 900, 925 and 925 ^{0}C.

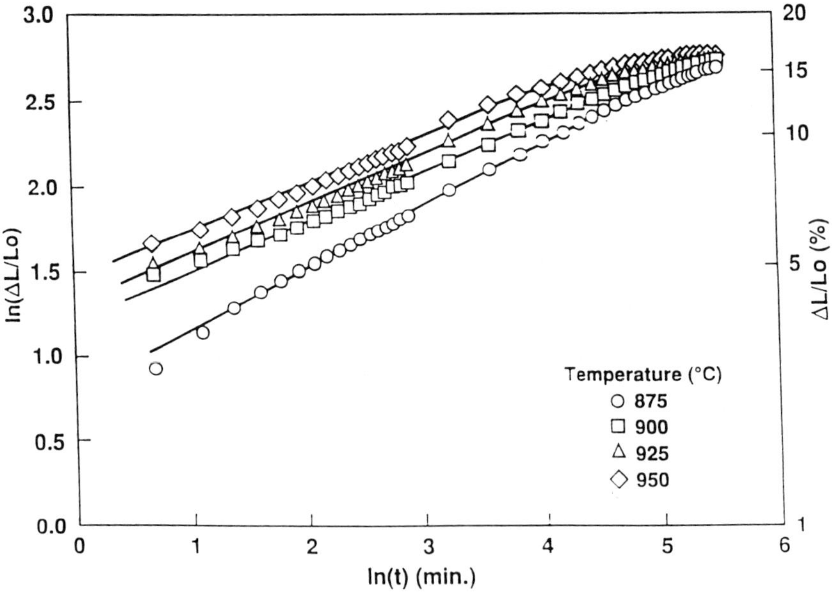

Figure 4. Isothermal shrinkage data of Fig. 3 plotted as $\ln(\Delta L/L_0)$ vs $\ln(t)$.

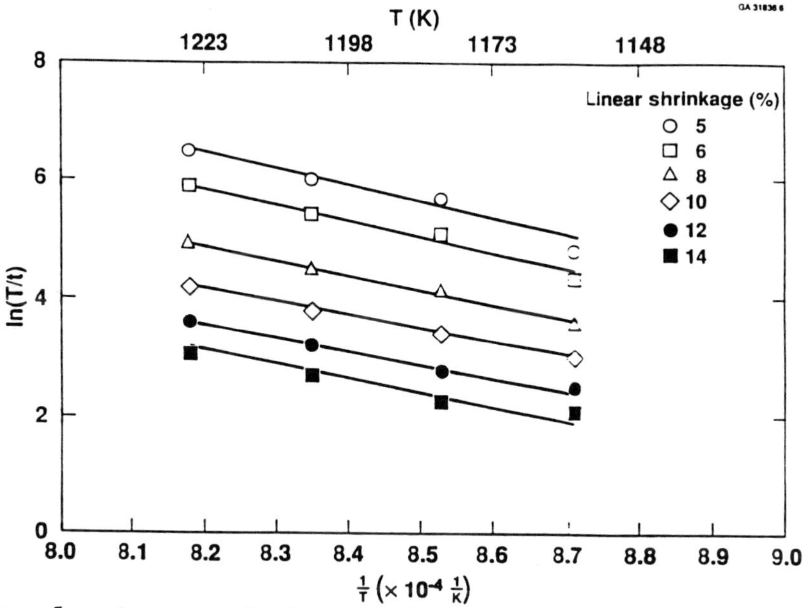

Figure 5. Apparent activation energy determination at various linear shrinkages for the shrinkage data obtained isothermally in Fig. 3.

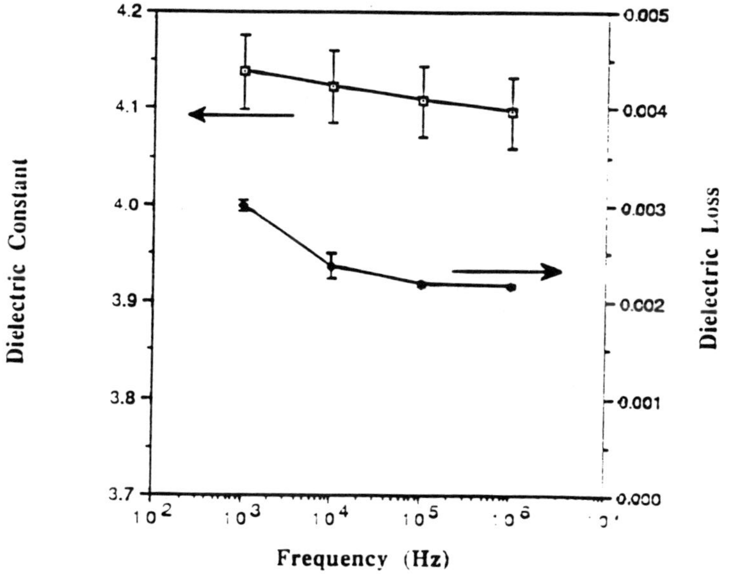

Figure 6. Plots of dielectric constant and loss as a function of frequency.

CORRELATION OF THE MECHANICAL TO THE THERMOMECHANICAL STRENGTH OF CERAMIC PACKAGES

A. Villani and H. Nguyen
Digital Equipment Corporation
77 Reed Road, MS: HLO2-1/L08
Hudson, Massachusetts 01749
UOG/Semiconductor Packaging

ABSTRACT

A correlation was conducted between the thermomechanical (thermal) and the mechanical strength of ceramic packages. The Acoustic Emission technique (AE) was used to detect the initiation of cracks in the ceramic packages as they are subjected to strength tests. A new mechanical and thermal strength testing technique was developed to evaluate ceramic strength. The temperature and force at which ceramic failure occurred was used in Finite Element Modeling to determine thermal and mechanical strength, respectively. Global models and submodels were used to obtain accurate strength results. Failure mode and Weibull statistical comparisons between these two techniques were also conducted. Very good correlation was obtained between the thermal strength and the mechanical strength of ceramic packages with the use of these techniques.

INTRODUCTION

The long term mechanical integrity of ceramic packages is a major concern to ensure the reliability of IC devices. The mechanical strength of ceramic packages can be evaluated by many popular methods, but their thermal strengths have not been mentioned much in the IC industry. The ceramic strength may be limited to the pass/fail qualification criterion in the temperature cycling and thermal shock tests.

To the extent authorized under the laws of the United States of America, all copyright interests in this publication are the property of The American Ceramic Society. Any duplication, reproduction, or republication of this publication or any part thereof, without the express written consent of The American Ceramic Society or fee paid to the Copyright Clearance Center, is prohibited.

A new Square Ring on Square Ring (SoS) ceramic strength testing technique was just developed. With this technique, a ceramic package is placed in the SoS fixture and a 'mechanical force' is applied to the ceramic package until the ceramic fails. The average force at 63% probability of failure is then used in FEM models of the ceramic package as it is subjected to SoS tests and strength results are obtained. Note that this is a 'mechanical strength' result. However, when the ceramic package is subjected to assembly and environmental conditions, it is subjected to not only mechanically induced stresses but also to thermal stresses. The term 'thermomechanical (thermal) strength' is used here to indicate the maximum stress at which the ceramic fails when it is subjected to thermal stresses.

Integrated circuit ceramic packaging is composed of many different materials which have various coefficient of thermal expansion (CTE). During assembly or environmental processes, the thermal stresses are generally caused by these CTE mismatches between the ceramic and other components, such as: the die, the lid, the heat sink, the module, and other materials, pending on the packaging scheme.

Although the strength of the ceramic can be quickly determined by mechanical means such as the SoS technique, the correlation between the mechanical and thermal strength is not known.

One source has pointed out that "an important advantage of the biaxial tests over the uniaxial tests is that the biaxial stresses attained during testing are closer to those stresses induced due to thermal stresses (1)." However, it was not known how 'close' thermally induced stresses were to mechanically induced stresses.

The approach of mechanically applying a force to failure to ceramics to determine its strength is very popular. Some of these techniques that are used with this approach are: 4PT-bend-testing(2), Ring-on-Ring(1), Hydraulic-Biaxial-Tension-Test-Apparatus (3)(4), Ball-on-Ring(1), and Piston-on-3-Balls(1). With these techniques the uniaxial and biaxial 'mechanical' strength of beams, plates, and disks are calculated. However, the issue of correlation between mechanical and thermal stresses is not addressed.

For Heat sink side tension, some correlation between mechanical and thermal stress was conducted at Digital by Mirman(5). Mirman used the force and temperature data obtained from SoS and AE results, in his analytical tools (ELD-SIR and ELD-RoR)(5), to obtain the mechanical and thermal 'maximum strengths' and the 'characteristic stresses'. The PIA and Shetty's approach (5) was used. These are accurate results when the ceramic packages are tested with the heat sink side in tension. However, with this technique, the stress concentration points at the cavity corners can not yet be accounted for. To account for stress concentration points, some FEM technique such as ANSYS 3D modeling and submodeling is needed.

To detect the point at which the ceramic package cracks, as it is subjected to thermal strength tests, the AE technique is used. This technique detects the temperature and the acoustic emission from cracks that are caused in ceramic packages during thermal strength tests.

This paper makes a first attempt at correlation between thermal and mechanical stresses. The 'maximum strength' of the cavity side of a ceramic package is tested with both SoS and by inducing thermal stresses.

OBJECTIVE

Compare the thermal to the mechanical strength of complete ceramic packages when they are subjected to cavity side strength tests. Use the following techniques:

a) SoS to determine the ultimate force to failure.
b) AE to determine the temperature and cracking signal.

Note that the thermal strength is calculated from the combined effect of a negative temperature cycle and the temperature variation between the heat sink, the ceramic, and the lid area.

MECHANICAL STRENGTH TEST APPARATUS

All tests were conducted with the SoS strength testing technique. A cross section of the strength testing apparatus is illustrated in Fig. 1. A ceramic package is placed in the SoS fixture and force is applied until the ceramic package fails.

TEMPERATURE DATA GATHERING PROCEDURE

First, some uncured epoxy is sandwiched between the heat sink side of the ceramic package and the 35.6 mm x 35.6 mm x 2.03 mm (1.4"x1.4"x.08") aluminum slug. The aluminum slug is held in place with a clip. Second, thermocouples are attached on the surface of the aluminum slug, on the ceramic (next to the aluminum slug), and on the package's lid. The ambient temperature of the oven is monitored with the oven's thermometer and with a separate thermocouple. Third, an acoustic emission device is assembled to the surface of the aluminum slug. Fourth, the package is placed in an oven, at an ambient temperature of 150°C, and the epoxy is cured for about 30 minutes. During the curing process, acoustic emission and all temperatures are monitored simultaneously. Sixth, the package/aluminum slug assembly are taken out of the oven. As the whole assembly cools, the temperature at which ceramic failure or acoustic emission occurs is recorded. This process was repeated for 10 packages. The curing temperatures (Ti), the temperatures at which the ceramic cracks (Tf), and the difference between these two temperatures (T) are illustrated in Table 1.

ACOUSTIC EMISSION

The Acoustic Emission technique has been used in the IC industry but its applications are relatively new. For thermal strength tests, a custom built transducer (AE sensor) which monitors the AE signal parameters such as amplitude, energy, and counts versus time (during the thermal stress testing) is attached to a ceramic package. The custom built sensors were developed for use with the thermal stress tests over a wide temperature range to avoid the piezoelectric self noise signal from sensors. The cracks in the ceramic are detected by analyzing the AE signals. Cracks were also inspected with microscopes, with SEM analysis, by cross-sectioning, and with C-scanning. Additional thermo-couples were also used to record the temperatures of the aluminum slug, the ceramic, the lid, and ambient.

RESULTS

Thermal and Mechanical Strength Results

After the complete global model of a typical ceramic package, illustrated in Fig. 2, is subjected to a negative temperature cycle, the thermal stresses illustrated in Fig. 3 are obtained. However, because of a singularity at the corner of the package, submodeling is necessary.

If a portion of the highly stressed region of the ceramic cavity corner is isolated and its model executed, the thermal stress contours of Fig. 4 is obtained. To insure mesh convergence, submodel 2 and submodel 3, illustrated in Fig. 5 and 6 were executed.

Thermal strength results from all global models and submodels are summarized in Fig. 7. This figure illustrates that when the second and third submodel is executed, the mesh has converged and the strength is accurate. This conclusion is also supported by the significant reduction in error of the energy norm.

The strength of the package was also evaluated with the SoS technique. The above global modeling and submodeling procedure was followed. Results from SoS strength tests are illustrated in Fig. 8. Once again this graph illustrates that when the second and third submodel is executed, the mesh has converged and the strength is accurate.

If the strength curve illustrated in Fig. 7 is superimposed on the strength curve illustrated in Fig. 8, Fig. 9 is obtained. Fig. 9 illustrates that there is excellent correlation between the thermal and mechanical strength of this ceramic package when it subjected to cavity side tension tests.

Stress from Temperature Variation

When the whole package is cooled from the curing temperature of the epoxy, 150°C, to room temperature, the ceramic fails at an average temperature of about 94°C. At this point, there is a temperature variation (or gradient) between the surface of the aluminum slug, the surface of the ceramic, and the surface of the lid as illustrated in Fig. 10. As a result of this temperature variation, the temperature isosurfaces of the ceramic package, illustrated in Fig. 11 are obtained. These isosurfaces may be input into a structural model and the thermal stresses are determined. Once again, to account for stress singularities, submodels were executed. The temperature variation stress results of the cavity corners of the ceramic package at 63.2% probability of failure is about 1 to 1.5 KPSI. This study illustrates that temperature variation stresses are not significant.

Weibull Statistics

"The measured strength and its variability are dependent on the distribution of flaws. A flaw becomes critical, causes failure, because of the combined effects of its size, shape, and orientation to the applied stress" (1). A number of fracture criteria are currently being used to analyze the impact of flaws's size distribution, their shape, and orientation(5); however, for the sake of simplicity and to make this memo brief, Weibull statistics will be used to analyze the thermal strength data.

If the ultimate force from SoS testing is used in ANSYS models, mechanical strength of each package is obtained. Likewise, if the temperatures at probe location 2 of Table 1 are used in the ANSYS models, the ultimate thermal strength of each ceramic package that was tested is obtained. These mechanical and thermal strength results are plotted in terms of Weibull statistics in Fig. 12. The thermal strength of a ceramic package that is subjected the cavity side tension at 63.2% probability of failure is 318 Pa (46.2 KPSI), with a Weibull Modulus of about 10; the mechanical strength, 322 Pa (46.7 KPSI) with a Weibul Modulus of 12.6. Note that the strength obtained with these two technique correlates quite well.

Failure Analysis

Failure modes of ceramic packages that are subjected to SoS tests are essentially the same as those subjected to thermal strength tests. The crack location and orientation is the same in both cases.

As illustrated in Fig. 13, the cracks of a ceramic package that is subjected to SoS (during cavity side strength tests) go through the cavity corners. For thermal strength tests, the failure modes are the same.

The acoustic energy vs time, output from the acoustic emission equipment, is illustrated in Fig. 14. The temperature has also been added to the x axis to simultaneously depict acoustic energy vs temperature.

When the ceramic and aluminum slug assembly are taken out of the oven at a temperature of about 150°C, the whole assembly begins to cool and simultaneously the epoxy begins to solidify. Consequently, because of CTE mismatch between the aluminum and the ceramic, thermal stress is applied to the ceramic by the aluminum slug. Note, aluminum has a CTE of 23.2(10)-6/°C and ceramic 6.2(10)-6/°C. When the whole assembly cools to temperature at about 90°C to 100°C, after about 1000 seconds, the thermal stress caused by the aluminum is significant enough to cause the first set of cracks in the cavity corners and the first loud acoustic emission is picked up. As the whole assembly cools, very small amounts of acoustic emission is picked up, implying that there is crack propagation. Then, at about 50°C to 60°C more major cracks are detected. At this stage, the initial cracks in the cavity corner have propagated to the point where the package's lid is stretched tight and the next set of cracks are caused next to the lid, the other stress concentration point. Finally, as the package cools from about 50°C to room temperature little or no additional cracks are picked up with acoustic emission.

This two phase failure mode process is also evident when the ceramic package is subjected to cavity side tension with the SoS technique. The cracks first occur in the cavity corners; then, if testing is continued, the lid stretches to a tight position and additional cracks occur right next to the lid. Both failure modes of both of these techniques are essentially the same.

CONCLUSIONS

When a complete ceramic package is subjected to cavity side tension, very good correlation is obtained between mechanical and thermal strength results. The Weibull slopes are very similar also.

Acoustic emission is a viable technique for determining the point at which the ceramic packages fail as they are subjected to assembly and environmental conditions.

The ceramic package failures obtained with the SoS technique are essentially the same failures as those obtained by applying thermal stresses to the package.

ACKNOWLEDGMENTS

The authors acknowledges with great appreciation and thanks to the help of the following people: Boris Mirman, Shih Hsu, Isa Bar-On, Joyce Hyde, Jamie Ide, John Sauber, Sheera Knecth, Steve Bosworth, and Ken Kulp who provided technical guidance and help with experimental studies.

REFERENCES

1. Sharon A. Mulligan. "Uniaxial and Biaxial Strength Evaluation of a Tape Cast Alumina." WPI Thesis. May, 1991

2. ASTM Designation: C1161 _ 90

3. D.K. Shetty, Et all. "Statistical Analysis of Size and Stress State Effects on the Strength of an Alumina Ceramic." ASTM STP 844. 1984, PP.57-80.

4. Antonio C. Rufin. "Statistical Prediction Models for Brittle Materials." AIAA Journal. Vol 22, No. 1, January 1984.

5. Boris Mirman "Translation of stress state into reliability term for single ceramic packages." Journal of Electronic Packaging. December, 1994.

Table 1. Thermal Strength of 339 PGA ceramic package when it is subjected to pin side tension.

Ti, INITIAL TEMPERATURE, °C @ Probe locatns 1, 2, 3,	Tf, TEMP. WHEN CERAMC CRCKS, °C @ Probe locatns 1, 2, 3,	(Ti-Tf) °C, @Probe Locations 1, 2, 3,	Strength, MPa/KPSI
156, 151, 151,	100, 88, 88,	56, 63, 63,	324/(47.0)
156, 151, 151,	100, 88, 88,	56, 63, 63,	324/(47.0)
156, 151, 151,	115, 100,108,	41, 51, 43,	262/(38.0)
156, 151, 151,	110, 95, 103,	46, 56, 48,	288/(41.8)
157, 152, 152,	103, 89, 96,	54, 63, 56,	324/(47.0)
157, 152, 152,	98, 85, 90,	59, 67, 62,	345/(50.0)
157, 151, 152,	115, 101,109,	42, 50, 43,	257/(37.3)
157, 151, 152,	102, 89, 95,	55, 62, 57,	319/(46.3)
157, 152, 152,	113, 95, 98,	44, 57, 54,	293/(42.5)
157, 152, 152,	104, 88, 91,	53, 64, 61,	329/(47.7)

Note:
T(ambient) = 158°C. (measured.)
Probe location 1: On the heat sink slug.
Probe location 2: On the ceramic, next to the heat sink.
Probe location 3: On the ceramic lid.

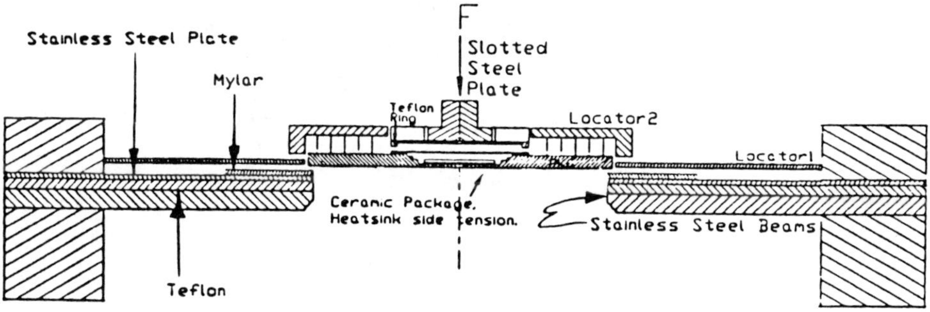

Fig. 1. Square ring on square ring (SoS) ceramic package strength testing apparatus. Cross section of apparatus.

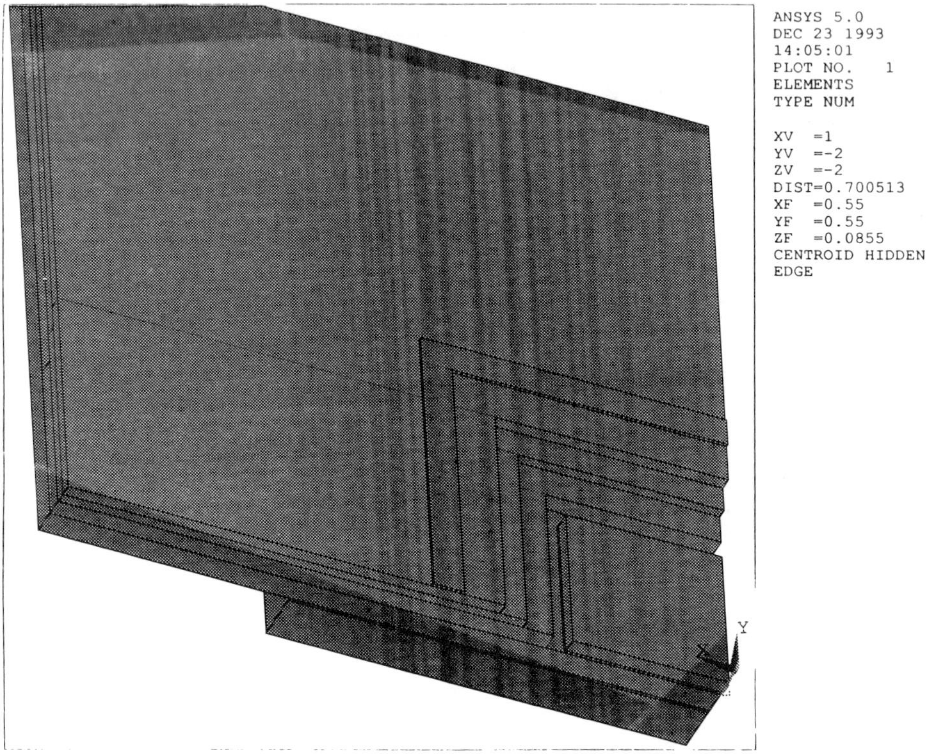

Fig. 2. Generic global model which will be subjected to thermal strains caused by a negative temperature cycle (Delta T = -100 C).

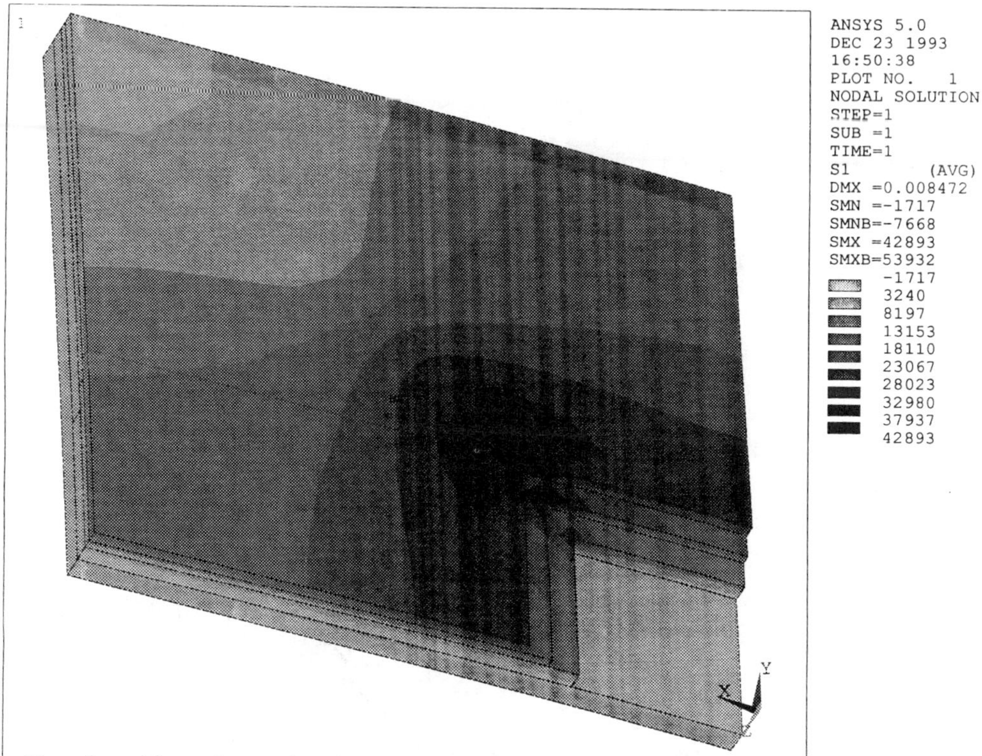

Fig. 3. Thermal strains in the ceramic caused by a negative temperature cycle (Delta T = -100 C).

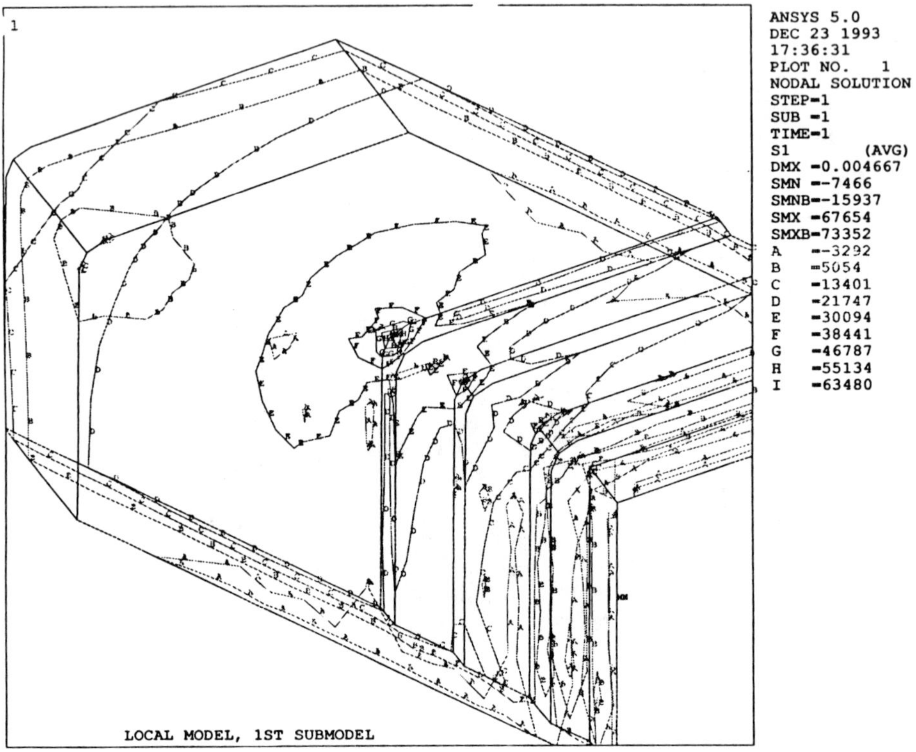

Fig. 4. Thermal strains in submodel 1 of the ceramic caused by a negative temperature cycle (Delta T = -100 C).

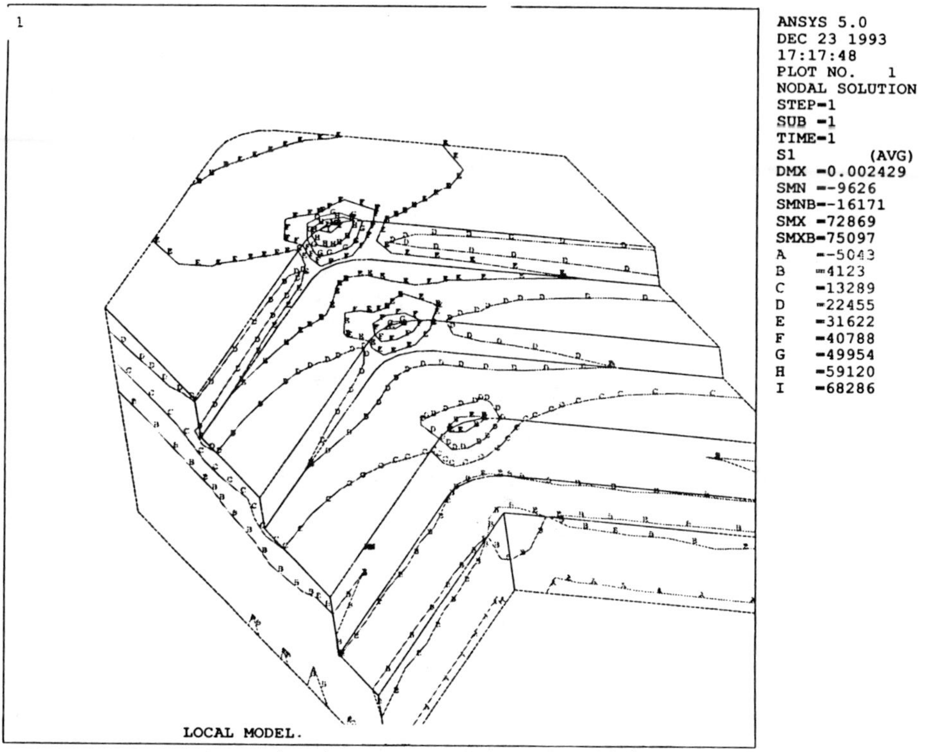

Fig. 5. Thermal strains in submodel 2 of the ceramic caused by a negative temperature cycle (Delta T = -100 C).

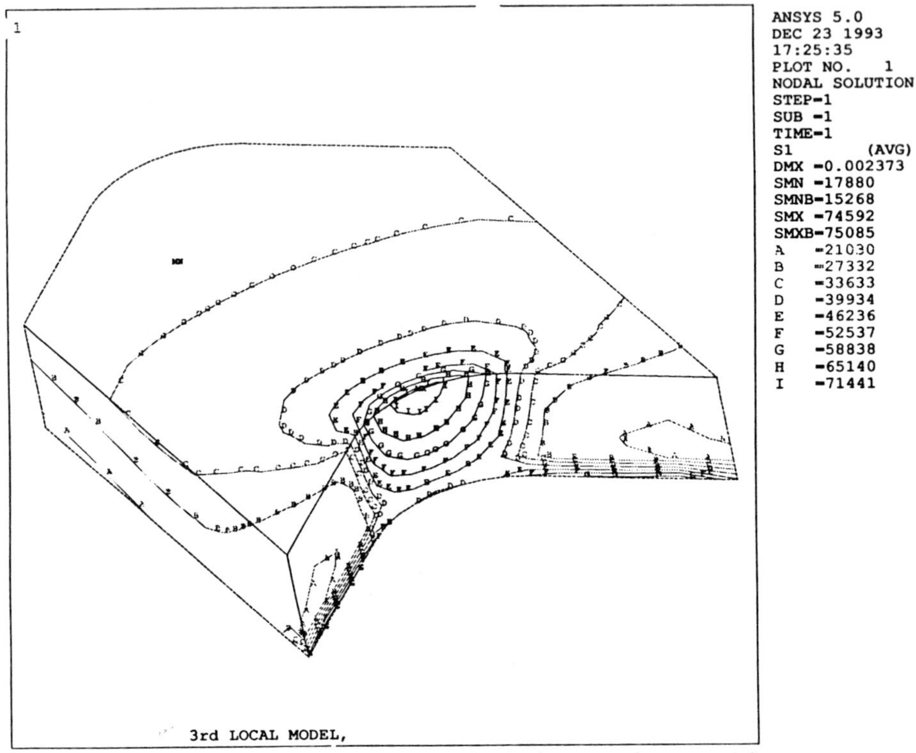

Fig. 6. Thermal strains in submodel 3 of the ceramic caused by a negative temperature cycle (Delta T = -100 C).

Fig. 7 Mesh convergence from a Global model to Submodel 3.

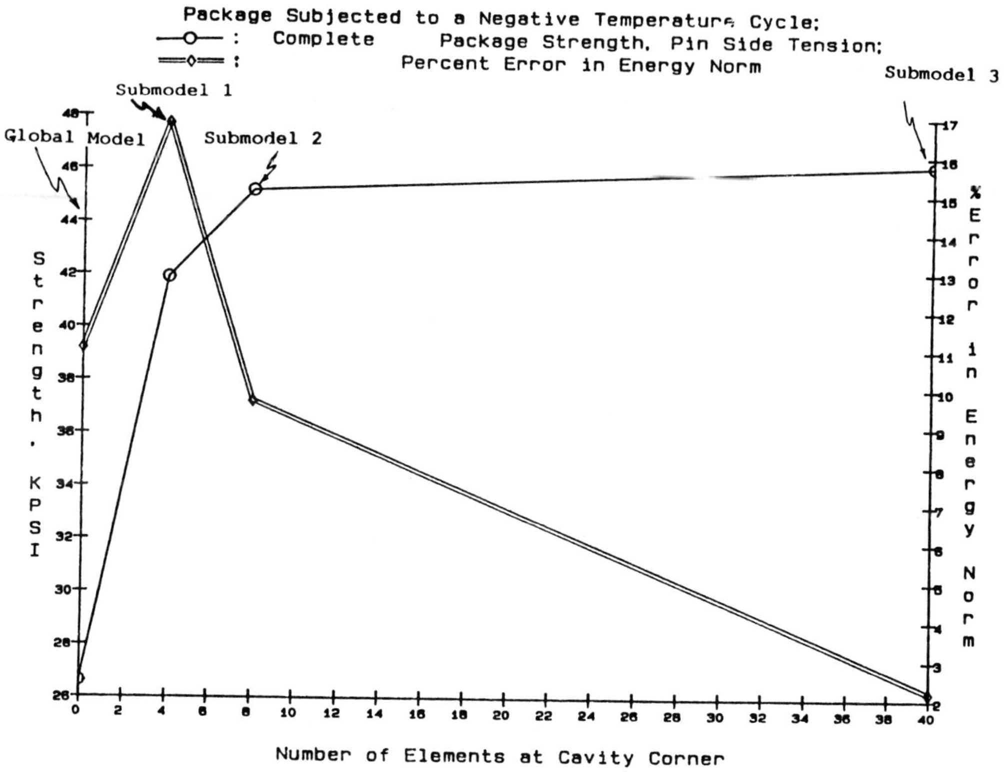

Fig. 8 Mesh convergence from a Global model to Submodel 3.

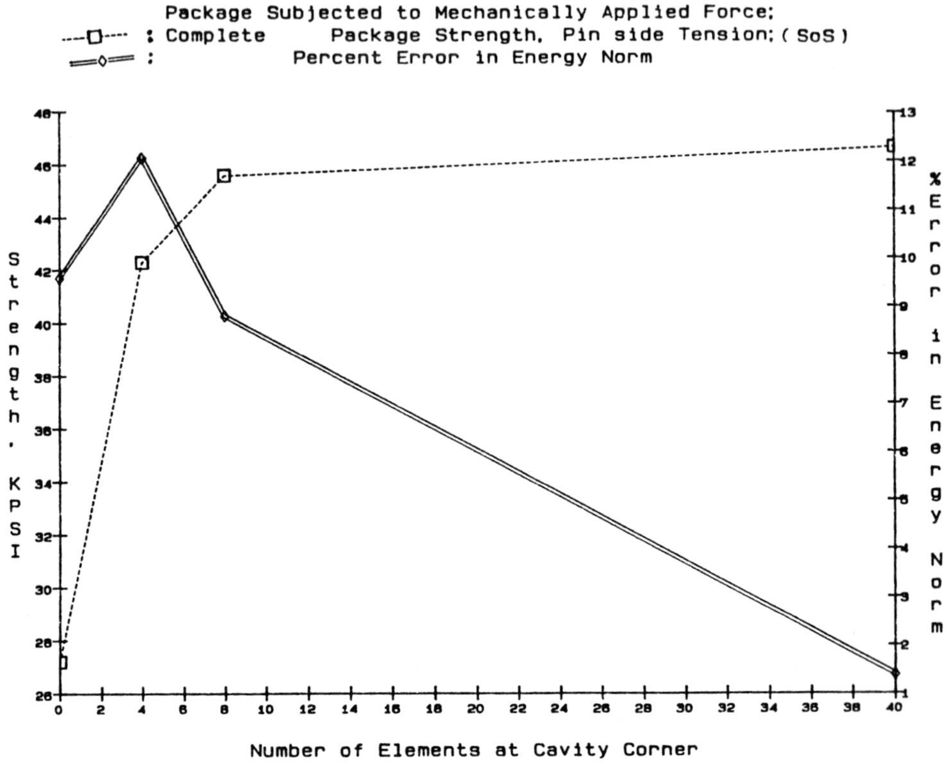

Fig. 9 Mesh convergence from a Global model to Submodel 3. Comparison between mechanical and thermal strength.

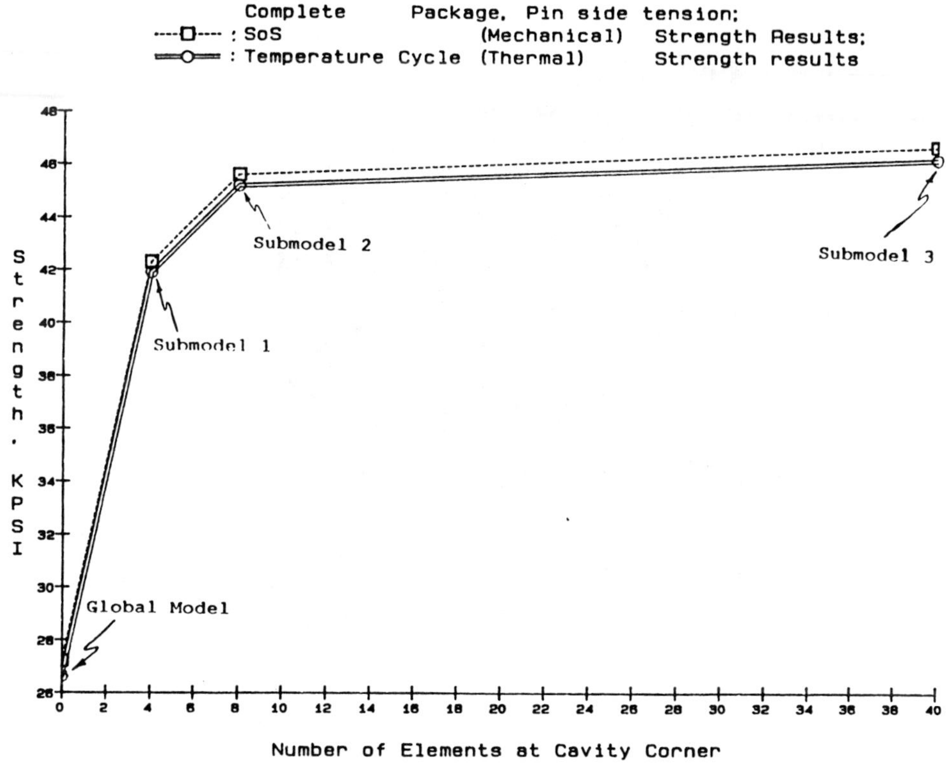

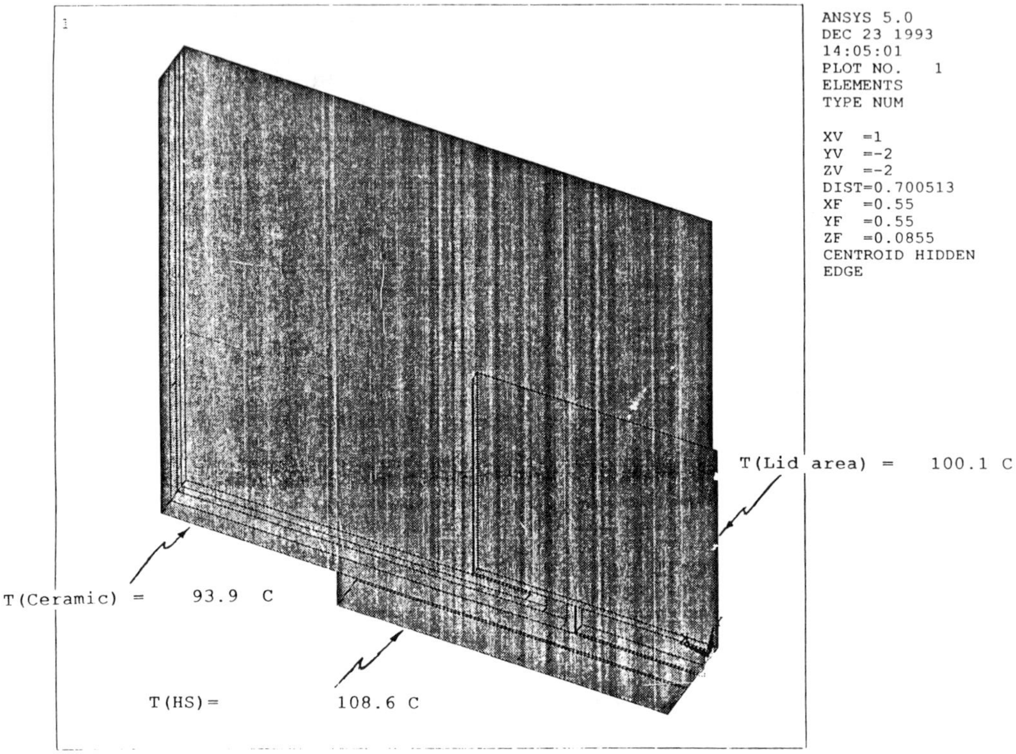

Fig. 10. Generic global model with temperature variations between the aluminum slug, the ceramic package, and the lid.

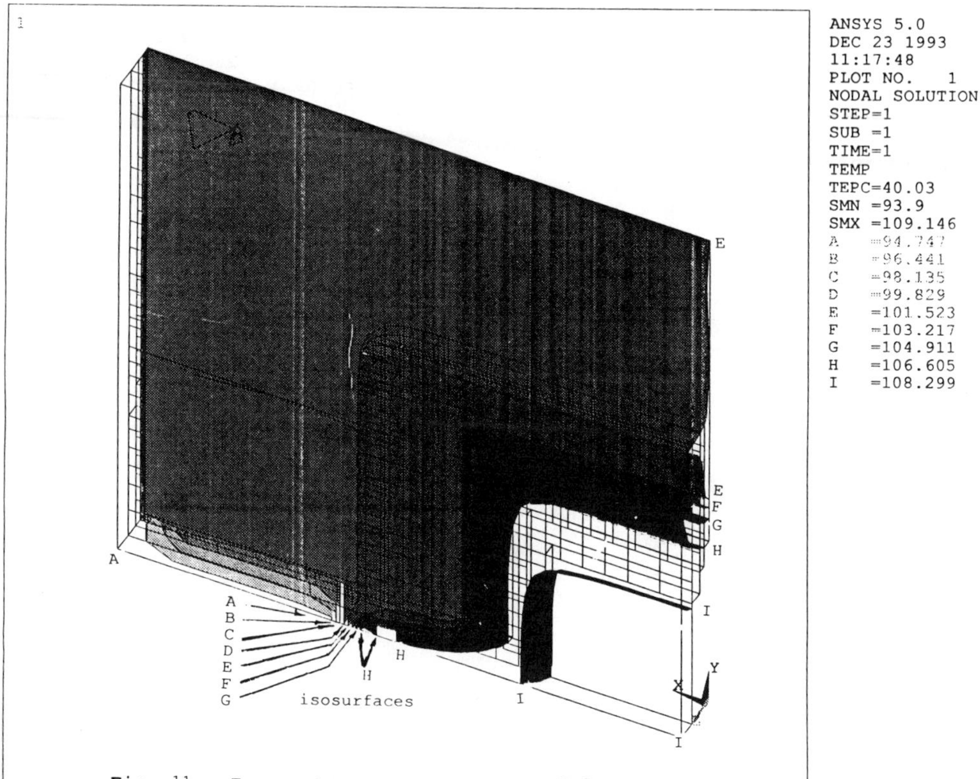

Fig. 11. Temperature Isosurfaces. (°C)

Fig. 12. Comparison of the thermal to the mechanical strength;
@ 63.2% Probability of failure of 339 PGA ceramic;
Thermal Strength: Beta= 9.992, Stress @ 63.2% = 46.27 KPSI;
Mech Strength : Beta= 12.62, Stress @ 63.2% = 46.68 KPSI

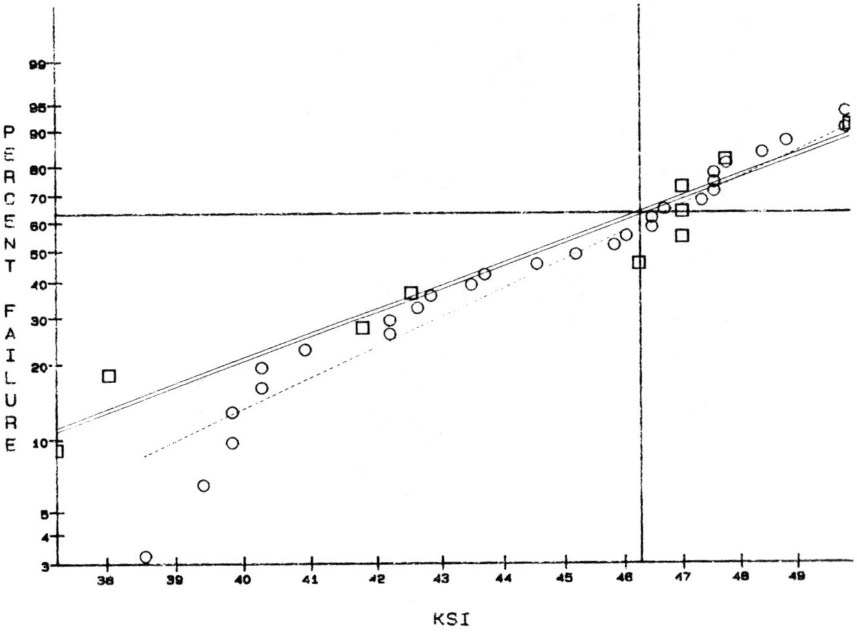

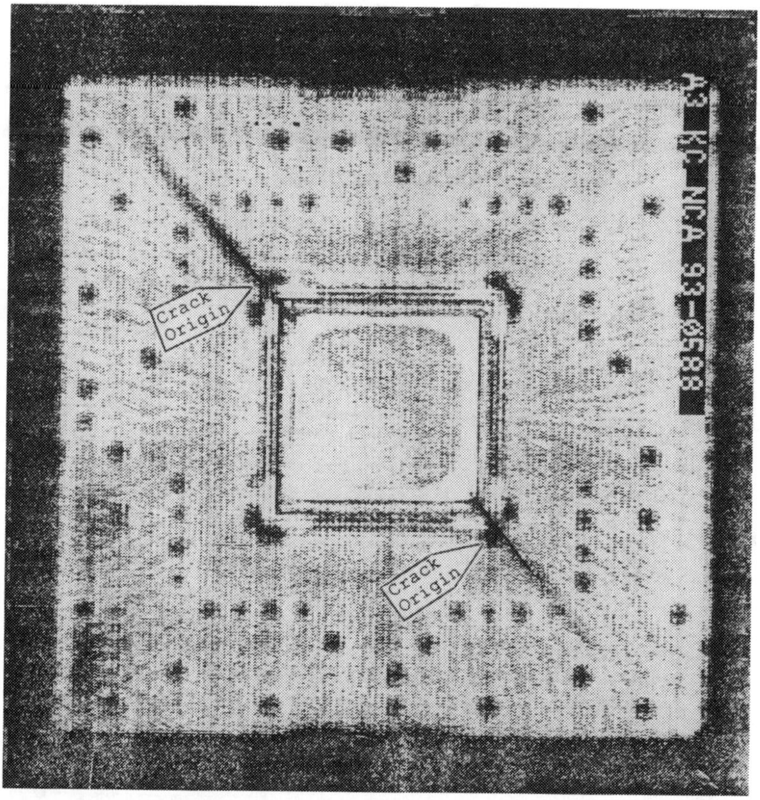

Fig. 13. **Crack pattern of a 339 PGA complete ceramic package as it is subjected to SoS pin side tension. Typical failure modes. Cracks originate at the cavity corners. C-scan image of crack at corner.**

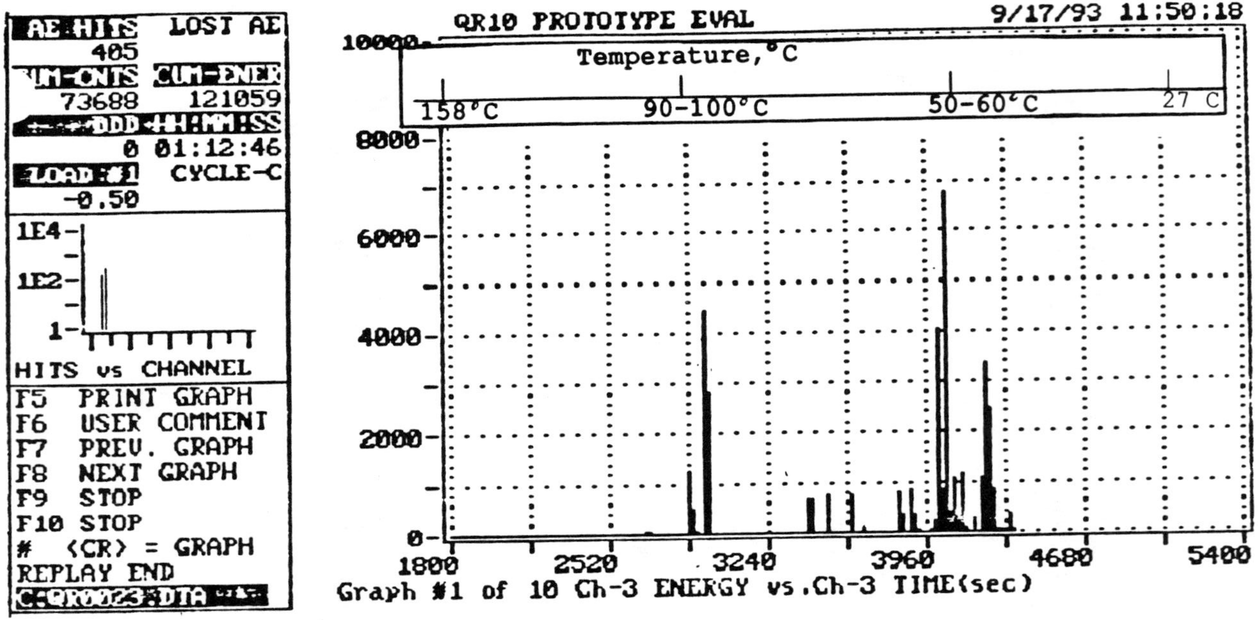

Fig. 14. Acoustic Energy (db) vs Time (sec) and Temperature (C).

Note:
- the cracks are first heard at T=90-100°C, t=2900 seconds
- crack propagation occurs between 90 and 60°C.
- more cracks heard at T=50-60°C, t=3900 seconds

POWERED THERMAL TRANSIENTS MODELING TO COMPARE AlN SUBSTITUTION FOR BeO

Gary M. Crosbie and Tsung-Yu Pan
Research Laboratory, Ford Motor Company, Dearborn, MI 48121-2053

ABSTRACT

Because of the potential reliability benefits of cooler circuits, highly thermally conducting materials are of interest in temperature-critical electronics. Because they have thermal conductivity (κ) higher than that of BeO, recently developed AlN substrate materials are of special interest.

Finite element analyses of thermal transients were run with higher and lower AlN thermal values substituted for those of BeO in a representative case: a heat spreader. As steady state conditions are approached, the 3-D finite element model ranks the two AlN's and BeO in a manner that is consistent with simple analytical estimates: a higher κ leads to a lower transistor temperature. Due to less expansion coefficient mismatch, normal and shear stresses in the solder layer under the die are found to be larger in magnitude for both AlN's than for the BeO.

Qualitatively, the lower "volume specific heat" ($\rho \cdot C_p$) of AlN might be expected to provide a faster heating ramp of the transistor than would be estimated from its thermal conductivity alone. Modeled temperature transients for power-on transients are virtually the same -- within 1 °C -- for the BeO and the higher-κ AlN.

INTRODUCTION

In automotive electronics, demands continue to expand for reliability and thermal performance. In particular, the underhood ambient temperature continues to increase, because requirements for more passenger space continues to lead to a reduction of available underhood volume and airflow. With these higher temperatures for powertrain control modules, higher thermal conductivity materials, such as AlN and BeO,[1] are of increasing interest to keep device temperatures inside at the same or lower temperatures as in earlier module designs.

To the extent authorized under the laws of the United States of America, all copyright interests in this publication are the property of The American Ceramic Society. Any duplication, reproduction, or republication of this publication or any part thereof, without the express written consent of The American Ceramic Society or fee paid to the Copyright Clearance Center, is prohibited.

For a particular geometry and value of generated heat, a higher thermal conductivity material should lead to a lower "ΔT" through the module stack under steady state conditions. Keusseyan et al.[2] and Hopkins et al.[3] previously reported numerical models comparing of older AlN's to BeO, without the temperature dependence of thermal conductivity.

In Figure 1, the temperature-dependent thermal conductivities of two aluminum nitrides are compared with BeO. The newer, low-oxygen AlN provide thermal conductivities that are higher than beryllium oxide (BeO), above room temperature, according to data provided by Tokuyama.[4] With the other AlN materials, the AlN advantage for thermal conductivity appears only at higher temperatures. For underhood temperature above 100°C, substitution of any of these AlN's should provide a lower "ΔT" at steady state. However, steady state conditions are only part of the requirement. Typical use includes powered transients. Further, reliability depends on control of stresses developed.

In this work, a comparison is made of the transient stress and thermal behaviors of thermal stacks modeled with BeO and the two AlN materials.

MODELING

The modeling simulation of different heat spreader materials was carried out with three-dimensional finite element analysis (FEA). A transistor stack was modeled using PATRAN and was analyzed with the ABAQUS general-purpose FEA program version 4.7-21. Both steady state and powered transient analyses were performed.

The transistor stack model consisted of layers of the power transistor, a solder layer, the ceramic heat spreader (BeO or AlN), another solder layer, a copper layer, and an aluminum baseplate. The dimensions for each of these layers are given in Table I. The model was constructed with 8-node brick heat-transfer elements (DC3D8) using a half-symmetric plane vertically through the middle of the stack.

The heat was generated in the transistor layer with 14.78 W constant power output after time zero. For the die dimensions given in Table I, this heating corresponds to 57.3 W/cm^2. To simulate the convection problem, the outer surfaces of the exterior elements were given appropriate heat transfer film coefficients with a 25°C ambient. In the interior of the model, the heat transfer was taken to be entirely by conduction, with no interfacial resistances. Temperature-dependent values for the thermal conductivity for the AlN's were taken from Tokuyama,[4] as shown in Figure 1. Designations used here are "SH-04" for SHAPAL™ SH-04 and "Sup-SH" for Super-SHAPAL™. "Values for temperature-dependent heat capacity for AlN and BeO were obtained from Touloukian and Buyco.[5]

Stress analyses were carried out by transferring the temperature results at each node to another finite element model with 8-node stress/displacement elements (C3D8). The new model was then analyzed again with ABAQUS. Both the normal and shear stresses were obtained for the solder layer between the heat spreader and the silicon die. Values are for the first FEA cell layer below the silicon die.

RESULTS AND DISCUSSION

Steady State Temperature Comparisons

An example of the steady-state temperature distribution is shown in Figure 2. This is a for a cross-section of the stack through the center of the silicon die. Under the silicon is a thin solder layer and then the heat spreader. In this case the heat spreader is the highest thermal conductivity AlN, designated "Sup-SH" -- as noted above, Super-SHAPAL™. The highest temperature overall is at the top center of the silicon and the highest solder temperature is in the center between the silicon and the heat spreader, at about 136°C. The inclusion of convection in the numerical calculation is reflected in the appearance that the silicon die corners are cooler. There is about a 9°C drop from the top of the silicon to the bottom of the heat spreader. As expected, the bulk of the temperature drop is in the thick aluminum baseplate.

The maximum steady-state transistor temperatures are in the sequence expected from the ranking of the 100°C-and-above thermal conductivities of Figure 1. The other heat spreader values values: The calculated values range from 139.52°C for the older "SH-04", through 138.95°C for the BeO, to 137.95° for "Sup-SH."

Solder Layer Stress Comparisons

The normal stresses are compared for the three heat spreader materials, in Figure 3. The normalized distance of "0" corresponds to the point where the temperature values are a maximum at the center. (Qualitatively similar stress results were reported previously by Pan and Pao[6] in discussion of a single heat spreader material.) Unlike the temperature case, the rankings here do not follow the thermal conductivity order. Indeed, the maximum stresses are found (in the outermost element along the edges of the silicon die) for the two AlN cases. This result is interpreted as a reflection of the thermal expansion coefficient differences, which lead to less expansion mismatch in the BeO case.

Similarly, comparison values for the shear stresses in the solder layer are shown in Figure 4. Here, the two AlN cases again overlap. The AlN values are larger in magnitude across the width of the solder under the silicon die, not just in the last element. Again, the expansion coefficient mismatch heat spreader materials are the basis for the differences.

Powered Transient Thermal Comparisons

As shown in Figure 5, the half-rise time at the transistor surface is about 20 sec for each of the three heat spreader materials examined. To aid in seeing the small differences among the three cases, the data are replotted as differences relative to BeO in Figure 6.

The AlN with the lower thermal conductivity ("SH-04") is 1.5°C above the "Sup-SH" from the first data point plotted (at about 10 seconds) out to steady state. This is interpreted as a result of the lower thermal conductivity with similar temperature dependencies of the two AlN's. It is noted that concerns for the conductivity measurement method may mask a stronger temperature dependence[7] for AlN than represented by values used here.

In Figures 5 and 6, there is little difference in transient temperatures between the two highest conductivity materials -- BeO and AlN:"Sup-SH" for the first 20 sec of the transient. In spite of the higher thermal conductivity of the "Sup-SH" AlN, the temperature at the transistor is as high for AlN as for the lower conductivity BeO, during this period.

In analysis of non-steady state heat flow, heat accumulation within an element is included in the numerical calculation. In particular, the lower volume-specific heat ($\rho \cdot C_p$) for AlN vs. BeO can be interpreted as a short-time offset that tends to lead to higher temperatures than otherwise during the early part of transients, in spite of the higher thermal conductivity of newer AlN. Of course, for slower heating rates, the contribution of the volume-specific heat decreases, eventually reaching zero at steady state.

In the present comparison of the two highest thermal conductivity materials -- BeO and AlN:"Sup-SH," a difference in transistor temperatures appears above 100°C. This difference increases continuously to about 1.0°C degree as steady-state is approached. This is consistent with the earlier discussion of the thermal conductivity comparison for steady-state.

CONCLUSIONS

As steady state conditions are approached, the 3-D finite element model ranks the two AlN's and BeO in a manner that is self-consistent with simple analytical estimates: higher thermal conductivity leads to a slightly lower transistor temperature.

Normal and shear stresses in the solder layer under the die are larger in magnitude for both the high and low thermal conductivity AlN's than for BeO, due to the greater expansion mismatch with the solder for the AlN's.

In power-on transients, modeled temperature increases are virtually the same (less than 1 °C) for the BeO and the higher thermal conductivity AlN, with the heat capacity effect showing only at earliest times.

This 3-D FEA transient analysis illustrates the opportunity to examine quantitatively the reliability-related materials choices for automotive electronics design.

ACKNOWLEDGMENT

The authors are grateful to W. L. Winterbottom and Ford Research Laboratory for provision of computing resources for the FEA calculations and for participation by one of us (G.C.) in the AlN Working Group.

REFERENCES

1. D. Suryanarayana, "Thermally Conductive Ceramics for Electronic Packaging," J. of Electronic Packaging (Trans. ASME), 111: 192-98 (1989).

2. R. L. Keusseyan, P. T. Goeller, and T. W. Whitehead, "Modelling of a New Packaging Concept for High Thermal Conductivity Applications," pp. 528-34 in Proceedings of the 3rd International SAMPE Electronics Conference, Los Angeles, California. (ISSN: 0891-0138)

3. D. C. Hopkins, S. H. Bhavnani, L. Tsai, and K. H. Dalal, "Numerical Modeling and Experimental Comparison of Copper Bonded AlN, Al_2O_3, and BeO Power Hybrid Structures," in Proceedings of 1992 Meeting of the International Electronics Packaging Society, Austin, Texas.

4. Tokuyama Soda Co., Ltd., "Technical Bulletin 8103: SHAPAL™ Translucent Aluminum Nitride Ceramic," San Mateo, California (1989).

5. Y. S. Touloukian and E. H. Buyco, "Thermophysical Properties of Matter: Vol. 5, Specific Heat -- Nonmetallic Solids," IFI/Plenum, New York, 1970.

6. T.-Y. Pan and Y.-H. Pao, "Deformation in Multilayer Stacked Assemblies," J. Electronic Packaging (Trans. ASME), 112: 30-34 (1990).

7. T. B. Jackson, K. Y. Donaldson, and D. P. H. Hasselman, "Temperature Dependence of the Thermal Diffusivity/Conductivity of Aluminum Nitride," J. Am. Ceram. Soc., 73 [8]: 2511-14 (1990).

Table I. Stack Dimensions for Finite Element Modeling

Layer	Function	Material	Length (mm)	Width (mm)	Thickness (mm)
1	Transistor	Silicon	5.08	5.08	0.25
2		Solder	7.62	11.6	0.025
3	Heat Spreader	BeO or AlN	7.62	11.6	0.635
4		Solder	7.62	11.6	0.086
5		Cu	7.62	11.6	0.695
6	Baseplate	Al	74.50	26.5	1.6

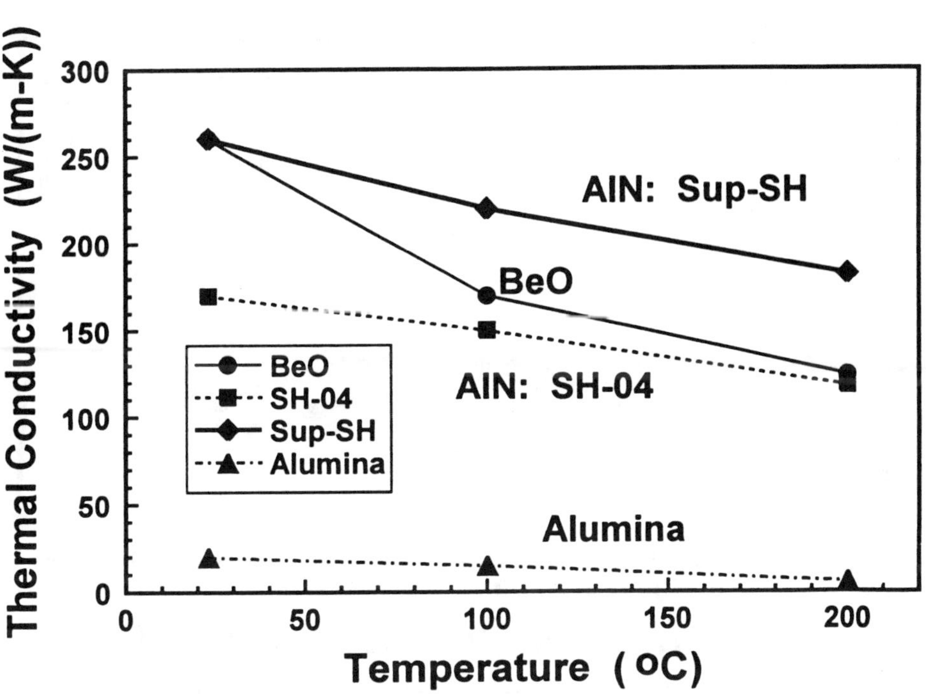

Fig. 1. Thermal conductivities of ceramic heat spreader materials.

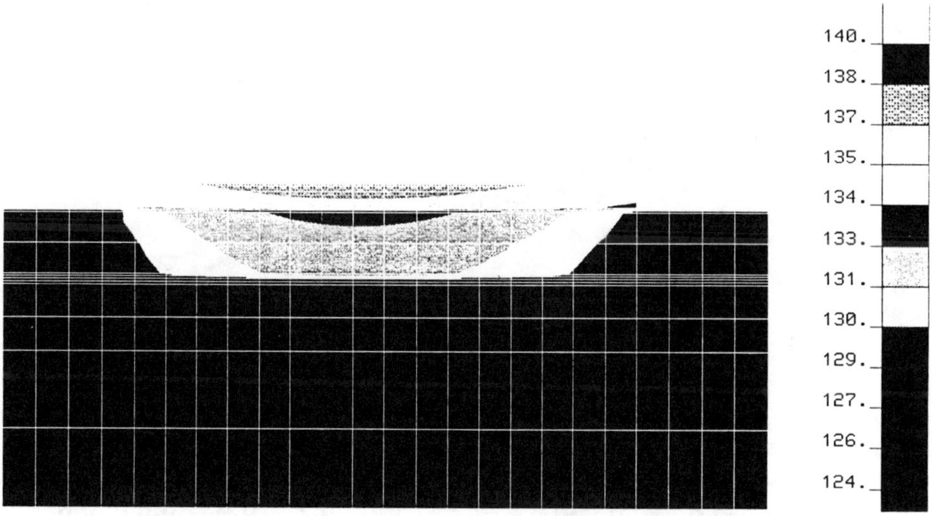

Fig. 2. Temperature distribution at steady state for a cross-section

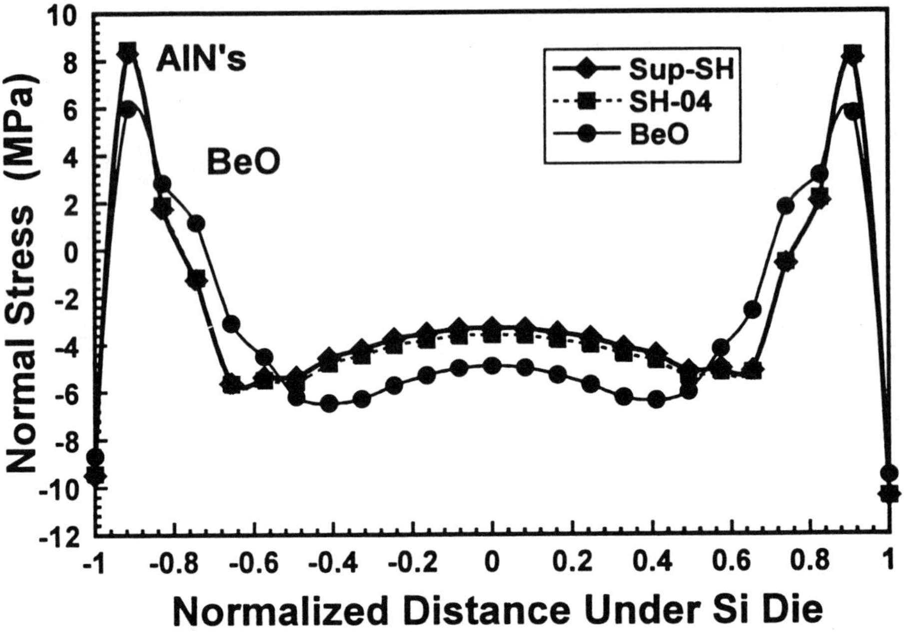

Fig. 3. Comparative normal stress levels in the solder layer under the die

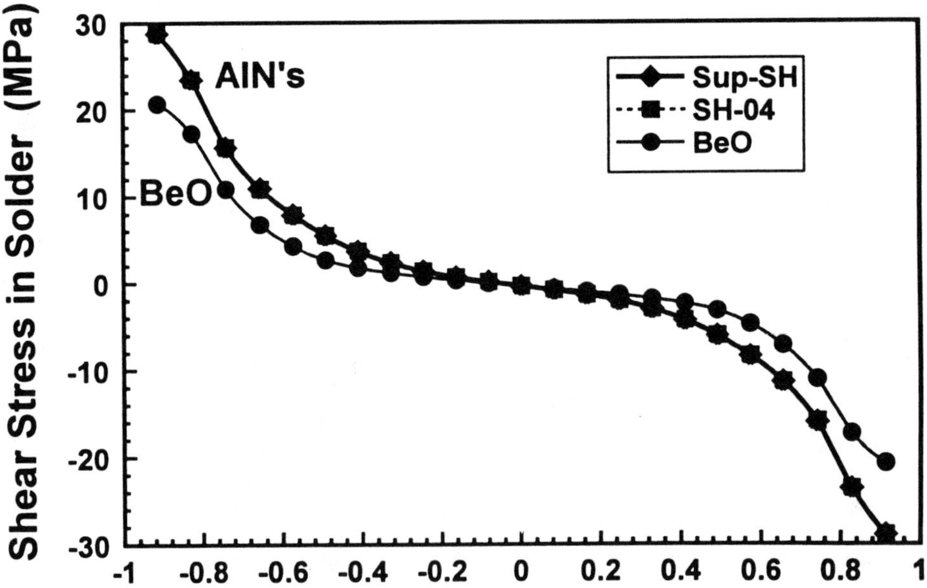

Fig. 4. Comparative shear stress levels in solder layer under the die

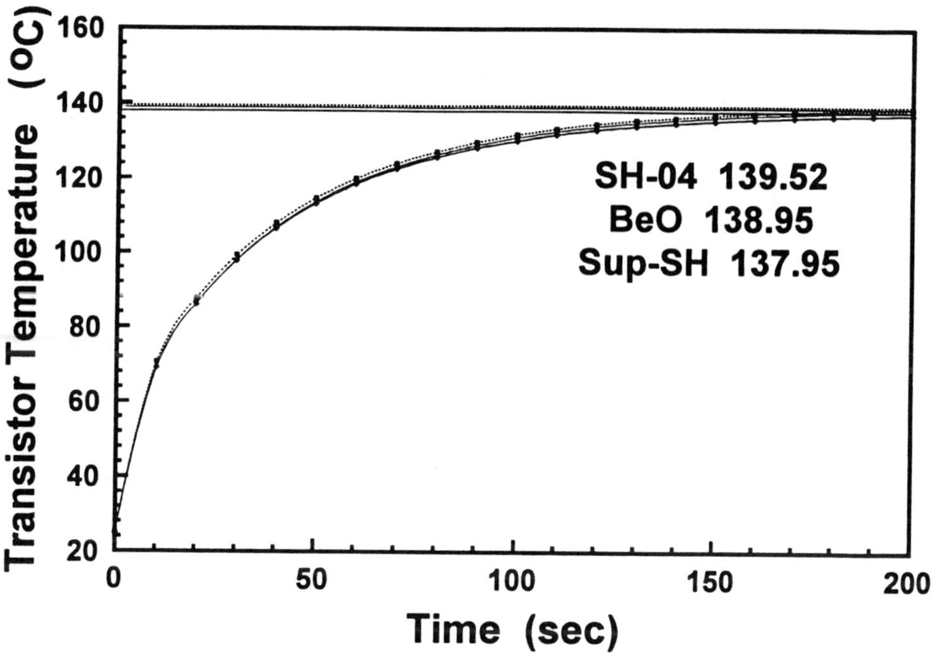

Fig. 5. Transistor temperature increase with time after power-up

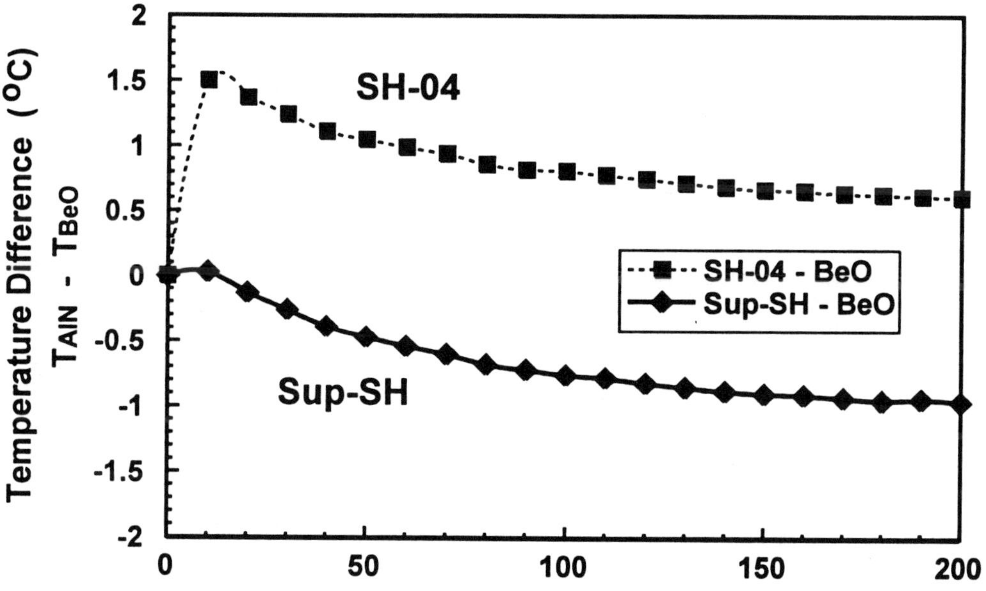

Fig. 6. Powered transient temperature differences relative to those of BeO

Pd Conductor for Thick Film Hydrogen Sensor

J. J. Felten
DuPont Electronics
14 T.W. Alexander Drive
Research Triangle Park NC 27709
Ph. 919/248-5688,FAX 919/248-5715

B. S. Hoffheins and R. J. Lauf
Oak Ridge National Laboratory*
P.O. Box 2008
Oak Ridge TN 37831-6002
Ph. 615/574-5620, FAX 615/576-2813

Cooperation between a materials developer and and sensor designers has resulted in a palladium conductor used ro design and build a new hydrogen sensor that has superior performance characteristics and is also inexpensive to manufacture. Material characteristics give it faster response time and greater thermal/mechanical stability. The thick film palladium conductor paste, which can be fired at 850°C-950°C, has provided device designers a practical conductor paste with which to produce the improved sensor. The conductor uses a high surface area Pd powder combined with a binder glass that is chemically very inert, which combination produces a porous conductor that has good adhesion and chemical resistance. The current sensor design consists of three or four thick film layers. Because of the flexibility of thick film techniques, the sensor element can be configured to any desired size and shape for specific instrument needs.

Key words: palladium conductor, dendritic powder, thick film, gas sensor, hydrogen sensor

Introduction

The uses for hydrogen sensors are numerous and increasing. Applications include military and aerospace vehicles, lead/acid battery charging systems, hydrogen furnaces, and nuclear reactors The ideal hydrogen sensor should be highly reliable, small, inexpensive, completely selective for hydrogen, and operable under a wide variety of atmospheres and temperatures. No commercially available sensor currently approaches this ideal. For example, gas chromatographs are selective and sensitive. However, they are large, expensive, and require frequent, expensive support and maintenance. Igniter systems are similarly bulky and limited in applicability.

Solubility of hydrogen in palladium is a well known phenomenon, even outside of the scientific community, as a result of the publicity surrounding the cold fusion controversy several years ago. When the surface of palladium or a palladium alloy is exposed to gaseous hydrogen, hydrogen molecules will dissociate and diffuse into the metal, ultimately reaching an equilibrium concentration that is a well-characterized function of the temperature and hydrogen partial pressure[1].Hydrogen solubility in palladium, hence ambient concentration, can be determined by measuring one or more of a number of properties of palladium, including electrical resistivity
[2-8]. The literature shows several attempts to use electrical resistivity as an indicator of the concentration of absorbed hydrogen in palladium, for example by Bendix in 1964 [2], by Sandia National Laboratory [9], and currently by Oak Ridge National Laboratory, a partner in this collaboration.

*Managed by Martin-Marietta Energy Systems, Inc., for the U.S. Department of Energy under contract DE-AC05-84OR1400.

To the extent authorized under the laws of the United States of America, all copyright interests in this publication are the property of The American Ceramic Society. Any duplication, reproduction, or republication of this publication or any part thereof, without the express written consent of The American Ceramic Society or fee paid to the Copyright Clearance Center, is prohibited.

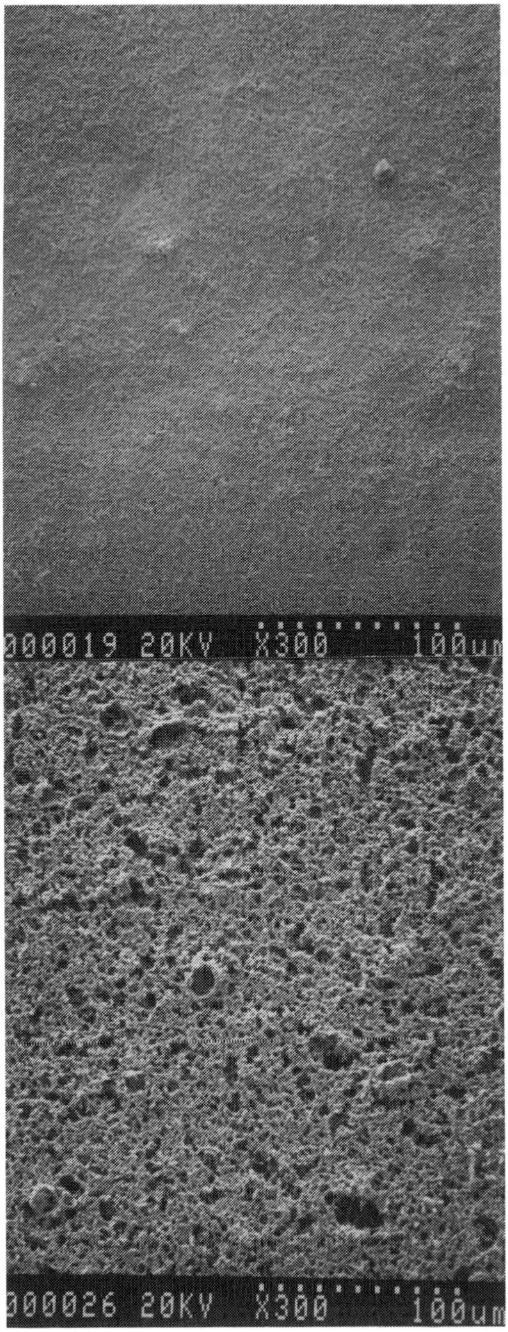

Design of Conductor Paste

When the possibility of formulating a special conductor for hydrogen sensors was first discussed,the sensor designer had already tested a series of Pd/Ag pastes to determine the necessary percentage of Pd necessary to achieve significant resistivity change in the presence of hydrogen. Since at least 73% Pd in a Pd/Ag sensor was found to be necessary for minimal response, it seemed logical to concentrate on compositions consisting entirely of Pd metal for the conductive phase. Palladium conductors have the potential to maximize the sensor's response to hydrogen. A small particle size, high surface area powder (0.1 micron dendritic powder with 10 meters/cm^2 surface area, was selected to achieve sintering at relatively low temperatures. Sintering to a void free fired structure was unnecessary and even undesirable, since a high degree of porosity and high surface area in the fired structure would aid in achieving rapid absorption of hydrogen into the Pd conductor. Photo 1 shows SEM views of the surface of the conductor, both fired and unfired, to show both green and fired structures of the Pd prints. For the binder, a glass of medium t_g was selected to permit firing at 850°C.[8] The glass also has excellent chemical resistance as measured by chemical etch tests of conductors. The paste formulation is listed in Table I, along with physical and electrical properties of the paste and fired conductor.

Photo I (on left): Dried (top) and fired(bottom) films of Pd conductor.

TABLE I: Pd Conductor Paste Formulation

Pd Powder		50%
Glass:	PbO	1.8%
	B_2O_3	0.1%
	SiO_2	1.4%
	Other	0.7%
Vehicle:	Ethyl Cellulose	5.9%
	Phthalate Ester	20.2%
	Terpineol Isomers	19.9%

TABLE II: Resistance of Printed Traces of Pd Conductor

Firing Temp	*R,Milliohm/Sq/mil*	*Thickness (microns)*
850o	140	7
950o	80	5

Fabrication of the Sensor Circuit

The sensor consists of three or four prints of thick film pastes used to construct a Wheatstone bridge type circuit [10-11]. A Pd/Ag paste (Dupont 6120) is used to print a pattern containing pads used to solder conductive leads to. The second print is the actual sensor layer discussed above consisting of the four elements of a Wheatstone bridge, with two elements to be passivated to form a reference for the elements actually exposed to hydrogen. The third material is printed to form the trimmable segments used to balance the Wheatstone bridge. Either a low ohm resistor (Dupont 1711) or a conductor (Dupont 6120) can be used. The fourth material, a passivation layer for the reference elements of the bridge, can be either a crystallizing multilayer dielectric (Dupont 9429) or a resistor encapsulant (Dupont 9137). Two prints of the passivation dielectrics were used to ensure protection of the reference legs of the bridge.

Characteristics of Thick Film Design

The thick film version of the hydrogen sensor is inexpensive to produce. It is small and lightweight, and uses little power. Many, if not most, applications will benefit from having available a device with these attributes. Particularly in applications requiring a large array of sensors, in consumer applications, especially to monitor the atmosphere near large batteies or to monitor for fuel leaks, low cost, weight and power consumption will be major factors. Low power operation invites uses in battery operated or portable applications.

Thick film sensors also will survive a hostile environment better than other technologies. The glass encapsulation is very impervious to steam, solvent vapors, acidic or caustic environments, or temperature extremes where organic encapsulants would fail. The thick film circuit is more robust than most designs using competing technologies.

Performance Evaluation of Sensors

Wheatstone Bridge Measurements

The sensor was placed in a small plastic chamber (410 cm^3) and connected to input power across opposing electrodes on the substrate while output voltage was measured across the remaining two electrodes (the standard Wheatstone bridge). The input voltage was nominally 9 vdc and for the initial tests the bridge was tested unbalanced. Sensor output in volts was measured at programmed intervals by an a/d interface in a IBM compatible personal computer. Although the test was nominally conducted at room temperature, the resistance of the sensor was low enough that at 9 V the bridge current elecates the sensor temperature slightly.

At the start of the test, readings wre taken for one minute in air. Then a mix of 4%H_2/ 96%Ar gas was passed through the sensor chamber for 5 minutes; at 6 minutes, the gas was turned off and the chamber was opened to ambient air to remove the test gas as quickly as possible. Readings continued for 5 minutes more This particular test cannot measure absolute sensor response time because the concentration of the test gas in the chamber is changing, taking a significant amount of time to effectively reach equilibrium. On the other hand, by removing the lid of the chamber, a much more rapid replacement with air is achieved because of rapid dissipation of the Ar/H_2 mixture, allowing a fairly accurate assessment of sensor recovery time to air.

This test proved useful as a basic measure of sensor operation. Typically, the sensor bridge output range was 100 mV for sensors with one active element and three passivated elements when exposed to a 4% H_2 concentration, and 200 mV for sensors with two actxive elements and two passivated ones. A typical set of test runs for a sensor with one active element is shown in Figure 1. The thick film sensor's response to hydrogen gas in its first exposure is usually the slowest. This effect has also been observed with initial tests of many of the thin film sensors. In subsequent exposures, response time decreases, possibly indicating the removal of adsorbed oxygen, moisture, other imporities, or residual palladium oxide from the firing process. In fact, during the first exposure to hydrogen, the active palladium regions of the as-fired thick film sensor visibly change color from dark gray to pale gray, suggesting the removal of oxides left by the firing process. This conclusion is also suggested by the fact that the surface of the fired palladium film changes from black to dark gray to pale gray as the firing temperature goes from 750°C to 850°C to 950°C when fired using a 30 minute profile in a belt furnace. For the sensor of Fig. 1, the time to reach 90% of the maximum output was 95 seconds for the first test and 40s for both the second and third tests. Again, it is assumed that the time constant of the chamber itself dominates the observed response time.

Effect of Temperature on Sensor

To verify that the palladium thick films followed classical behavior in the presence of hydrogen gas, the sensor was exposwed to hydrogen at different sensor temperatures, as indicated in Figure 2. The sensor was heated indirectly, and an S-type thermocouple measured surface temperature. The resistance of one palladium element was monitored by a four point measurement. The element was first exposed to4%H_2/96%Ar for 10 minutes and then to air for 10 minutes. Gas flow throughout the test was maintained at 8 cm3/s (1 scfh). As seen in the figure, the initial resistance of the palladium element encreases with temperature. The TCR derived from these experimental measurements is $43x10^{-4}$, close to the TCR for pure palladium, $38x10^{-4}$. This suggests that the palladium component of the film composition dominated the resistor behavior. The element responded faster to hydrogen at higher temperatures. Similarly, recovery time was also faster at higher temperatures. The magnitude of the response decreases with increasing temperatures, as one would expect based on thermodynamic considerations.

Effect of Hydrogen Concentration

The resistance of the same sensor element was measured for three concentrations of hydrogen gas. Figure 3 shows the responses of a Palladium element heated to 280°C.The change in resistivity between 1% and 4% hydrogen is essentially linear.

Future Formulation work

Future work may adapt a high resolution patterning technique such as making the paste photopatternable, or diffusion patternable. Such techniques will permit greater sensitivity in the sensor by increasing the usable conductor pathlength in a given area, and lower power requirements. Sensor response time may also be speeded up by using narrower conductor lines patterned with advanced patterning techniques. Reducing print thickness to the minumum consistent with good yields will reduce the materials cost of the sensors. Input from sensor designers and fabricators will provide guidance as to the effect of fabrication variables on paste design.

Conclusions

1. A thick film materials system for fabrication of hydrogen sensors has many attractive features for the sensor designer and fabricator. Light weight, low power consumption, and rugged construction are important advantages thick film materials contribute to sensor design and performance.

2. Sensors fabricated with thick film Pd have a TCR essentially equal to that of pure Pd. Temperature effects are essentially compensated by measuring the difference between passivated and unpassivated legs in the Wheatstone bridge configuration.

3. Sensor response is linear over the range from 1% H_2 to 4%. As temperature increases, sensor reaction time decreases but sensitivity also decreases.

4. High resolution patterning techniques may be advantageous to enhance the performance of the thick film Pd conductor. These patterning techniques may also permit thinner conductgor traces to be deposited. Response time and sen sitivity are expected to be enhanced, and materials costs are expected to be reduced.

References

1. A. Lewis, *The Palladium Hydrogen System*, Academic Press, New York (1967).

2. A. Michaels, *Design, Development, and Prototype Fabrication of an Area Hydrogen Detector*,
NAS85282, Bendix Corp., Southfield MS (1964)

3. I. Lundström, "Hydrogen-Sensitive MOS Structures, Part 1: Principles and Applications," *Sens. Actuators*, **2**, 403-426 (1981/1982)

4. I. Lundström, "Hydrogen-Sensitive MOS Structures, Part 1: Principles and Applications," *Sens. Actuators*, **2**, 105-138 (1981/1982)

5. L.M. Lechuga, A. Calle, D. Golmayo, and F. Briones, "Hydrogen Sensor Based on a Pt/GaAs Schottky Diode," *Sens. Actuators B*, **4**, 515-518 (1991).

6. J. Godwin and R.V. Klint, "Hydrogen Detector System," U.S. Patent 3,927,555 (1975).

7. J. Shaver, "Bimetal Strip Hydrogen Gas Detectors," *Rev. Sci. Instr.*, **40**, 901-905 (1969).

8. A. Butler, "Optical Fiber Hydrogen Sensor," *Appl. Phys, Lett,* **45**, 1007-1009 (1984).

9. R.C. Hughes, R.J. Buss, S.W. Reynolds, M.W. Genkins, and J.L. Rodriques, Wide Range H2 Sensor Using Catalytic Alloys, SAND-92-2382C, Sandia National Laboratory, Albuquerque, NM (1992).

10. B.S. Hoffheins and R.J. Lauf, "Thick Film Hydrogen Sensor," U.S. Patent application filed Dec. 19, 1993.

11. J.J. Felten, "Palladium Thick Film Conductor," U.S. Patent application filed Dec. 19, 1993.

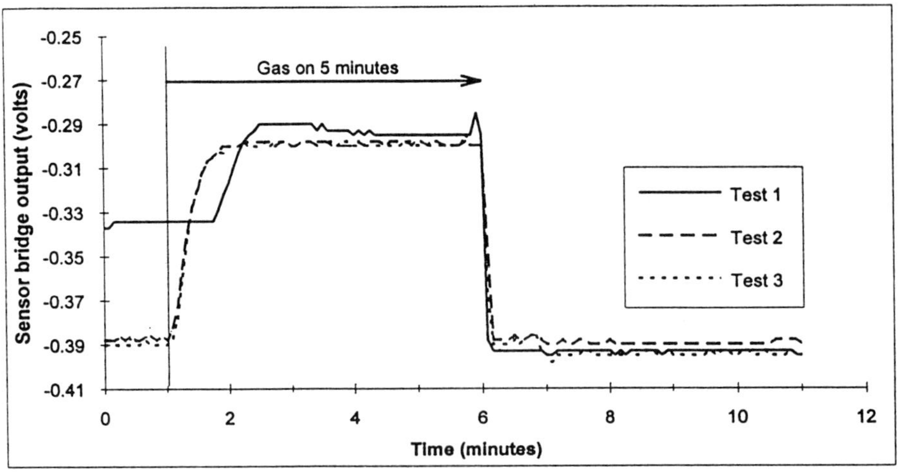

Figure I: Typical set of runs exposing sensor to hydrogen. The first test with a sensor always shows less voltage change.

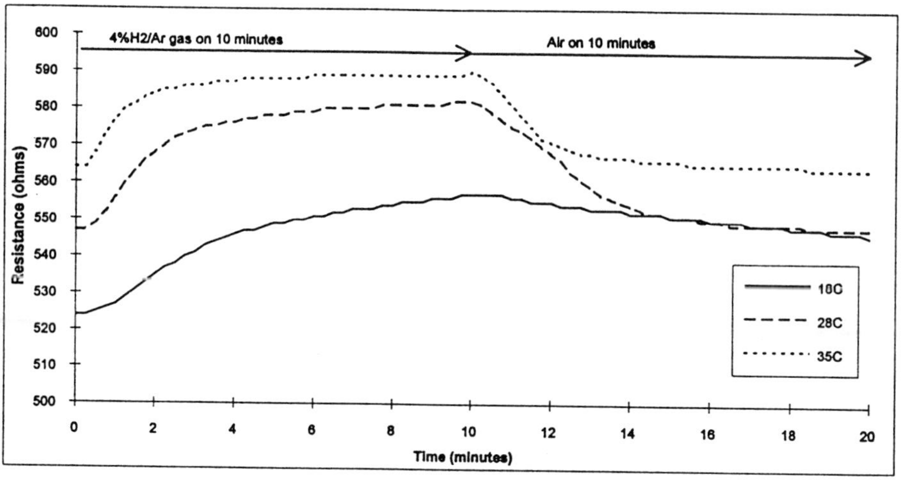

Figure II: The effect of temperature on Pd element response.

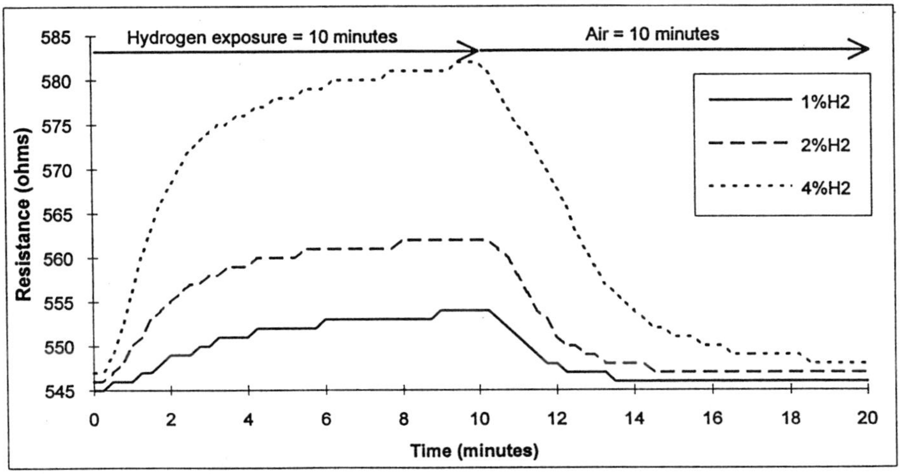

Figure III: The effect of Hydrogen Concentration on Element Resistance.

Ultrafine Alumina Coated Silicon Carbide Particles for Alumina-Silicon Carbide Nanocomposites

K. G. K. Warrier, U. S. Hareesh,
G. M. Anilkumar and A. D. Damodaran,
Regional Research Laboratory, Trivandrum-695019, India

Rolf Janssen,
Technical University of Hamburg-Harburg, Germany

Aluminium oxide, toughened by fine and uniform dispersion of silicon carbide particles has been found to possess interesting high temperature properties of high toughness and strength [1-3]. Recent reports suggest that the finer the silicon carbide size, the better would be the fracture tougheness. By additon of as few as 5 vol% submicron SiC particles strengthened Al_2O_3 from 350 MPa to 1 GPa, with tougheness as high as 4.7 Mpa $m^{1/2}$ [4-5]. The mechanism of such high extent of fracture toughness has been investigated to be due to crack deflection and microcracking introduced by thermal expansion mismatch between particles and matrix grains, although the real contributions from the silicon carbide nanoparticles are still under investigation [6]. In all these cases the primary requirement for most effective composite is the fine, uniform size of the particles and their homogeneous dispersion in the alumina matrix [7]. Usual methods adopted for the preparation of composites such as physical mixing of alumina and silicon carbide particles often result in localised agglomerations and inhomogeneity making the composite inferior in properties.

Homogeneous precipitation of aluminium hydroxide to coat the silicon carbide particles dispersed in an aluminium nitrate solution has been reported earlier to result in coated silicon carbide powders as a precursor for composite [7,8]. More recently mullite precoated SiC particles have been reported by plasma spray technique in view of the advantage that mullite has near identical thermal expansion coefficient with SiC [9]. Such precoated silicon carbide particulates will result in uniform dispersion in the alumina matrix and also prevent to some extent the oxidation properties of silicon carbide powder. However, conditions for arriving at uniform, thin, adherent coatings of alumina on silicon carbide are important to achieve this goal, for which appropriate sensitisation techniques are required. Hence an

To the extent authorized under the laws of the United States of America, all copyright interests in this publication are the property of The American Ceramic Society. Any duplication, reproduction, or republication of this publication or any part thereof, without the express written consent of The American Ceramic Society or fee paid to the Copyright Clearance Center, is prohibited.

attempt has been made in this study to sensitise silicon carbide particles of size ~ 0.3μm by heating and washing with hydrofluoric acid for short time, dispersing in aluminium nitrate solution in water and slowly precipitate the aluminium hydroxide monohydrate (Boehmite AlOOH) as a gel. Coated particles are then separated, heated to ~ 600°C to alumina coated silicon carbide and characterised. The uniform dispersion of silicon carbide in aluminium nitrate solution and the gelation were also followed by optical microscopy coupled with an image analysis system.

Experimental Procedure

As received SiC powder (0.27μm) was heat treated at 250°C for 2 h by washing with very dilute 1% HF to remove possible traces of silica associated with the powder . The HF treated powder was then washed thoroughly with double distilled water, finally with acetone and dried.

For developing boehmite gel coating on the SiC particles 1 gm sensitised SiC powder each was dispersed in 1 litre water at pH 2.5 and experiments were performed under conditions provided in Table I. 1 : 1 ammonium hydroxide was used for hydrolysis of aluminium nitrate. A drop of the dispersion of SiC before and after coating was observed on a glass slide under an optical microscope coupled with an image analysis system.

Table I

Concentration of Al : nitrate solution	Wt. of SiC (g)	Temp. °C	pH of hydrolysis	$SiC:Al_2O_3$ [by wt]
0. 01 M	1	90	7.5	1 : 0.1275
0. 05 M	1	90	7.5	1 : 0.6377
0. 1 M	1	90	7.5	1 : 1. 275
0. 5 M	1	90	7.5	1 : 6. 377

$Al(NO_3)_3 . 9H_2O$ solution was added to SiC suspension and the mixture was heated to 90°C with constant stirring. 1 : 1 NH_4OH was added dropwise till pH reached 7. 5. The coated SiC particles were removed from the mixture, dried under humidity conditions and later calcined at 600 °C for 3 h to get alumina coated SiC powders. These powders were characterised by XRD and Scanning Electron Microscopy.

Results and Discussion

Sensitisation techniques like heat treatment at 250 °C and washing with dilute HF acid produced activation sites at the SiC surface. Slow hydrolysis of $Al(NO_3)_3 \cdot 9H_2O$ solution produced very fine particles of boehmite over the SiC particles kept in constant dispersion at a pH less than 3.5 based on zeta potential measurement . As the hydrolysis progressed all the SiC particles got coated and finally settled once the hydrolysis was over at pH 7.5. The temperature was maintained at 90 °C throughout the hydrolysis and this considerably reduced the chances of agglomeration of SiC particles. The coated particles were first washed and then centrifuged so as to remove all the excess boehmite, if remained, from the suspension. The coated particles were dried in a humdity oven at 45 °C and 60 -70 % relative humidity for 24 h. The calcination of the coated powders were done in flowing nitrogen at a very slow rate of 1°C per minute to ensure crack free surfaces.

The XRD patterns of the SiC particles before and after coating from different solutions containing varying aluminium nitrate concentrations presented in Figure 1 indicates that the characteristic peaks for SiC decreases as the coating progresses. This decrease is gradual and indicates the progress of coating. The concentration of 0.5M solution is sufficient to completely cover the surfaces of SiC particles used in the present investigation.

The dispersion of SiC particles at $pH < 3.5$ as observed in the suspension under optical microscope coupled with an image analysis system with an effective magnification of 10000 indicates an average particle size of 0.3 μm (Fig. 2).

The SEM pictures presented in Fig. 3 distinctly indicate the coated particles. The basic size of the SiC particles before coating has been 0.3 to 0.4 μm with sharp edges. (Fig. 3a). The coated ones indicate that the edges are less sharp and there is partly agglomeration of SiC particles under the present experimental conditions. The platelets of boehmite distributed in small quantities around the coated particle agglomerates are shown in Fig. 3b where the coating solution had a concentration of 0.5 M. One of the isolated particles in Fig. 3b is presented at a higher magnification in Fig. 3c where a uniformity of coating including the edges is highlighted. Infact this concentration appears to be very high for the type of powders used for coating . On calcination at 600 °C the surface morphology shows coating on the particles although agglomeration caused in the coating process is still maintained to some extent (Fig. 3d). However, as the coating solution is reduced to 0.1 M there is considerable improvement in the agglomeration tendency and morphology of the coated SiC particles. Compared to the microstructure of the dispersion presented in Fig. 3e it appears that the agglomeration has been controlled only to limited extent during hydrolysis since effective dispersion was obtained in the starting aluminium nitrate solution. However by further controlling the rate of hydrolysis it should be possible to bring down the extent of agglomeration further.

Conclusion

The present study indicates the possibility of obtaining fully dispersed suspension of fine SiC particles of average particle size < 0.3μm in aluminium nitrate solution and also to obtain uniform coating by its controlled hydrolysis. A concentration of < 0.1 M aluminium nitrate and a temperature of 90°C would be optimum to obtain thin coating of boehmite which can be further calcined at 600 °C under very slow rates to result in thin alumina coated SiC powders. These powders can be effective precursors for alumina-silicon carbide nanocomposites.

References

1. K. Niihara and A. Nakahira " Strengthening of Oxide Ceramics by SiC and Si_3N_4 Dispersions" pp 919 -26, *Proceedings of the Third International Symposium on Ceramic Materials & Composites for Engines. The American Ceramic Society*, Westerville, OH (1988)

2. K. Niihara and A. Nakahira " Particulate Strengthened Oxide Nanocomposite " pp 637-64 , *Advanced Structural Inorganic Composites*, edited by P. Vincenzini, Elsevier, Trieste, Italy, (1990)

3. K. Niihara , " New Design Concept of Structural Ceramics-Ceramic Nanocomposite " The Centennial Issue of the Ceramic Society of Japan, *J. Ceram. Soc. Jpn.*, 99 (10) ,974-82 (1991)

4. L. C. Stearns, J. Zhao and M. P. Harmer, " Processing and Microstructure Development in Alumina-Silicon Carbide Nanocomposites" *J. Eur. Ceram. Soc.* , 10, 473-77 (1992)

5. J. Zhao, L. C. Stearns, M. P. Harmer, H. M. Chan, G. A. Miller and R. F. Cook " Mechanical Behaviour of Alumina-Silicon Carbide Nanocomposites" *J. Am. Ceram. Soc.*, 76(2), 503 -10 (1993)

6. A. Mark Thompson, H. M. Chan, M. P. Harmer, R. F. Cook "Crack Healing and Stress Relaxation in Alumina-Silicon Carbide Nanocomposites" *J. Am. Ceram. Soc.* 78 (3) 567-71 (1995)

7.Ching-Li-Hu and M. N. Rahaman " SiC Whisker Reinforced Al_2O_3 Composites by Free Sintering of Coated Powders " *J. Am. Ceram. Soc.* 76 (10) 2549-54 (1993)

8. H. Nakamura and A. Kato " Effect of Crystallisation of Alumina Hydrate in Preparation of Alumina Coated Composite Particles " *Ceram. Int.* 18 , 201-206 (1992)

9. K. N. Lee, R. A. Miller and N. S. Jacobson " New Generation of Plasma Sprayed Mullite Coatings on Silicon Carbide" *J. Am. Ceram Soc.* 78 (3) 705-710 (1995)

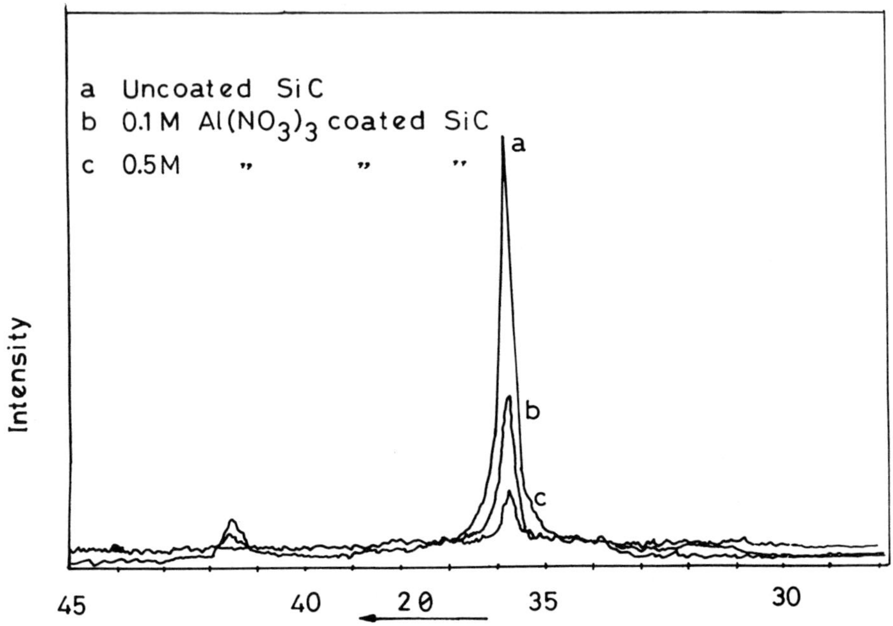

Fig. 4

Fig. 1 XRD patterns showing decrease in characteristic peaks of β-SiC after coating.

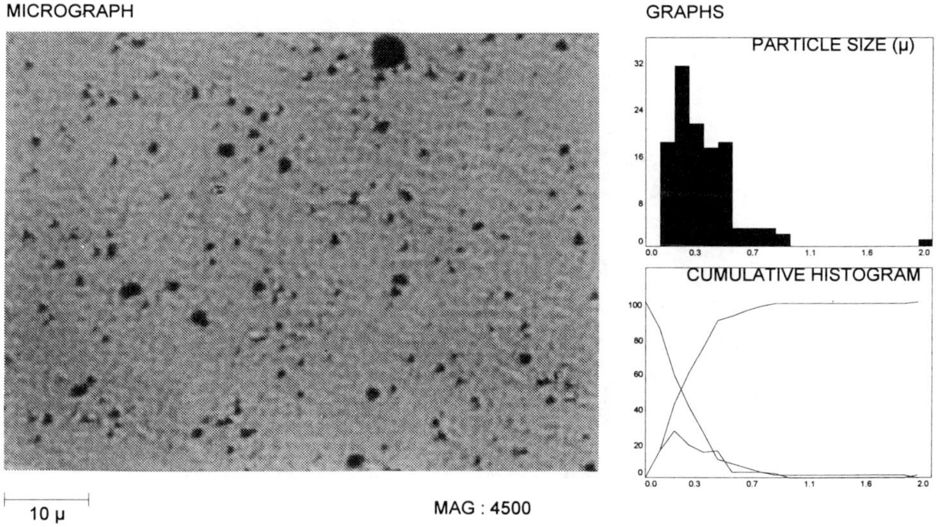

Fig. 2 Image analysis showing dispersion of fine SiC in Aluminium nitrate solution.

Fig. 3a SEM picture showing uncoated SiC particles

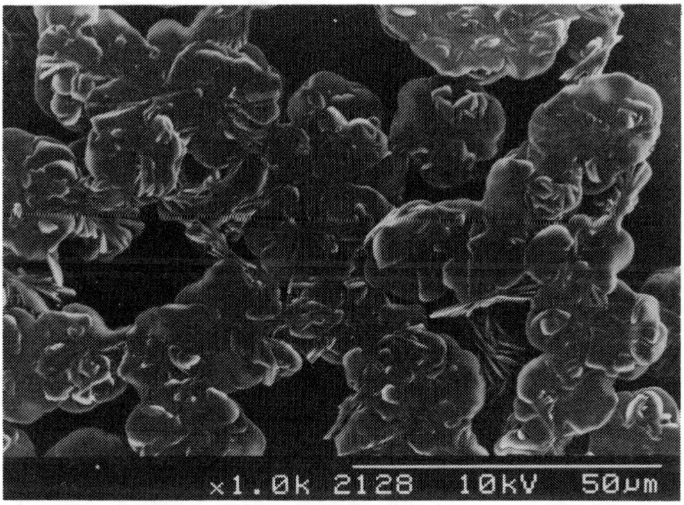

3b

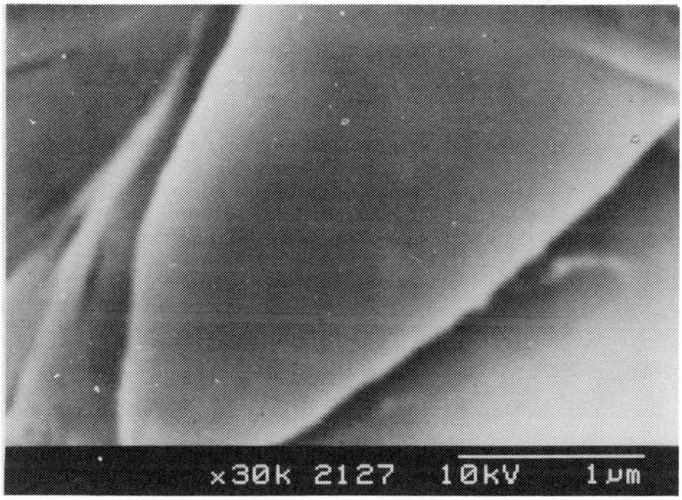

3c

Fig. 3b SEM picture showing Boehmite coated SiC
lower magnification
3c) Fig. 3b at higher magnification

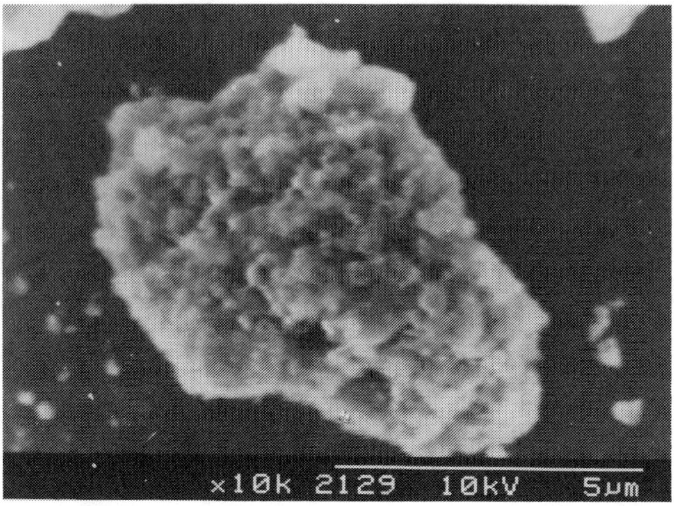

3d

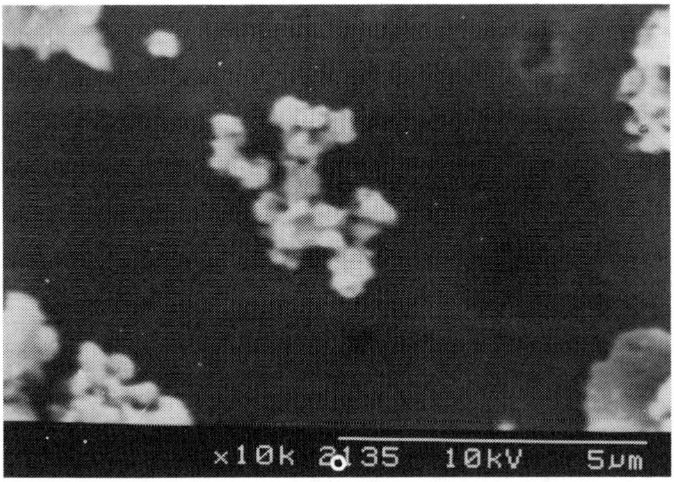

3e

Fig. 3d SEM picture showing Alumina coated SiC
0.5 M Aluminium nitrate coated SiC

Fig. 3e 0.1 M Aluminium nitrate coated SiC

NUMERICALLY CALCULATED EFFECTIVE PERMITTIVITY OF DIELECTRIC COMPOSITES

M.E. Azimi and P.K. Ghosh
Department of Electrical and Computer Engineering
Syracuse University
Link Hall,
Syracuse, NY 13244-1240

ABSTRACT

Modeling of the electrical properties of multi-component materials provides a thorough understanding of the nature of the resultant properties. In this paper we will focus on the dielectric response of the mixture. Many researchers have worked in this area after the pioneering works of Maxwell-Garnett and Rayleigh. Most of the reported methods are analytical in nature and are often either for a specific composition or for a specific inclusion shape. In our work we are developing a numerical approach with considerable flexibility to cover a wide range of multicomponent dielectric composites. The analysis starts with the development of integral equations representing the electrostatic potential distribution inside the composite. These equations are converted to approximate matrix equations using method of moments and then solved numerically. Theoretical results will be presented along with the calculated values using well known analytical approach.

INTRODUCTION

Calculation of the effective permittivity of multicomponent ceramic dielectric has been reported by many researchers. But most of the reported calculation are analytical and deals with spherical and some with elliptical inclusions [1-5]. Also, the discrepancies between the analytical and experimental results were discussed in terms of the component's shape [2,4] and interfacial effects [5].

The ceramic structures, in reality, are made of many dielectrics. In addition, dielectric inclusions could be of arbitrary shape and orientation. Thus, most of the

To the extent authorized under the laws of the United States of America, all copyright interests in this publication are the property of The American Ceramic Society. Any duplication, reproduction, or republication of this publication or any part thereof, without the express written consent of The American Ceramic Society or fee paid to the Copyright Clearance Center, is prohibited.

existing model can only simulate a specific or a simplified version of the actual multicomponent structure. In this paper we show the application of a numerical method which can estimate the effective permittivity of real world multicomponent ceramic dielectric such as dielectric with arbitrary inclusions. In this method we first estimate the polarization of the inclusions which depends on the inclusion's shape and their dielectric properties. Then, using a distribution function for these inclusions the effected permittivity of the composite is calculated.

NUMERICAL APPROACH

Consider a structure, as shown in Fig. 1, where an arbitrarily oriented lossless dielectric (permittivity ε_1) of arbitrary shape is embedded in a lossless homogeneous dielectric (permittivity ε_2). This composite structure is within another dielectric medium (ε_3). To simplify we will assume that this media is free space $(\varepsilon_3=1)$. To calculate the polarization of this structure we first write the integral equations to model the electrostatic potential distribution within the above mentioned structure (Fig. 1). Detail of the formulation is given by the authors in reference 6.

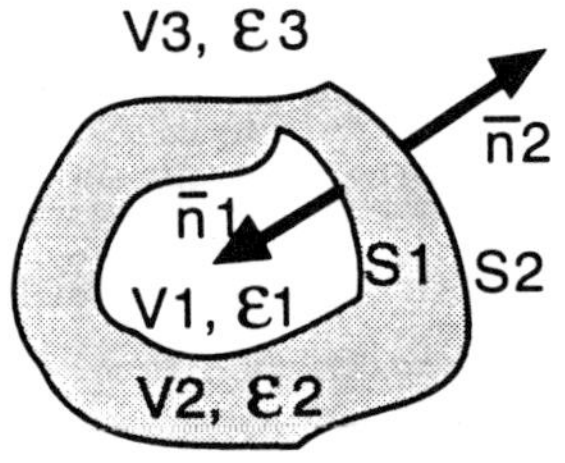

Fig. 1 A general composite for theoretical discussion. Vectors $\overline{n}_1$ and $\overline{n}_2$ are the unit vectors normal to boundary surfaces S_1 and S_2.

Let ψ (P_{si}) be the potential at point P_{si} due to certain distribution of charges in free space. Then, if ϕ_1 and ϕ_2 are the potentials on the surfaces S_1 and S_2, one can write the following two equations describing the surface potential distribution on S_1 and S_2

$$\psi(P_{S1}) = \frac{\varepsilon_1+\varepsilon_2}{2}\phi_1 + \frac{\varepsilon_2-\varepsilon_1}{4\pi}\int_{S1}\phi_1\frac{\partial}{\partial_n}\left(\frac{1}{r}\right)ds + \frac{\varepsilon_2-\varepsilon_3}{4\pi}\int_{S2}\phi_2\frac{\partial}{\partial_n}\left(\frac{1}{r}\right)ds \tag{1}$$

and,

$$\psi(P_{S2}) = \frac{\varepsilon_2+\varepsilon_3}{2}\phi_2 + \frac{\varepsilon_2-\varepsilon_1}{4\pi}\int_{S1}\phi_1\frac{\partial}{\partial_n}\left(\frac{1}{r}\right)ds + \frac{\varepsilon_2-\varepsilon_3}{4\pi}\int_{S2}\phi_2\frac{\partial}{\partial_n}\left(\frac{1}{r}\right)ds \tag{2}$$

where r is the distance between point P_{si} and the elemental area ds on S1 or S2. Note that the normal to the surface S2 is outward and the normal to the surface S1 is inward. The above two integral equations were then converted to approximate matrix equations using the method of moments [7]. The resultant matrix equations were then solved numerically to calculate potentials at various points on S_1 and S_2. The structural symmetry can sometimes be used to reduce the number of linear equations and thus the computation time.

Next, the dipole moment $\overline{P}$ of a homogeneous dielectric body with permittivity ε_{eff} embedded in free space can be expressed using a surface potential distribution by

$$\overline{P} = \varepsilon_0(\varepsilon_{eff}-1) \int_S \phi_0(p)\overline{n}_p \, ds \tag{3}$$

where "S" represents the surface enclosing the said dielectric body and $\overline{n}_p$ is the normal vector at point p on "S".
The effective dipole moment for the composite structure under consideration can then be formulated as follows:

$$\overline{P} = \varepsilon_0 \left\{ \frac{\varepsilon_2-\varepsilon_3}{\varepsilon_3} \int_{S2} \phi_2 (p)\overline{n}_p ds + \frac{\varepsilon_2-\varepsilon_1}{\varepsilon_3} \int_{S1} \phi_1 (p)\overline{n}_p ds \right\} \tag{4}$$

Next, consider a composite structure with multiple inclusions as shown in fig. 2. These inclusions are arbitrarily shaped and randomly oriented. Under an applied field the local electric field (E_ℓ) on each inclusion is related to the average electric field (E_{av}) in the composite by [4,9].

$$E_\ell = E_{av} + \frac{1}{\varepsilon_a} \overline{\overline{L}} \cdot \langle \overline{P} \rangle \tag{5}$$

where ε_a is the apparent permittivity "seen" by each inclusion within the composite structure. The depolarization dyadic $\overline{\overline{L}}$ is a symmetric dyadic which depends only on the inclusion's shape [8]. The average polarization $\langle \overline{P} \rangle$ is defined as

$$\langle \overline{P} \rangle = \int_{4\pi} \omega_0(\Omega) \, \overline{P} \, (\Omega) \tag{6}$$

where Ω represents three angle parameter which describe orientation of any inclusion and $\omega_0(\Omega)$ is the orientation distribution function of inclusions.

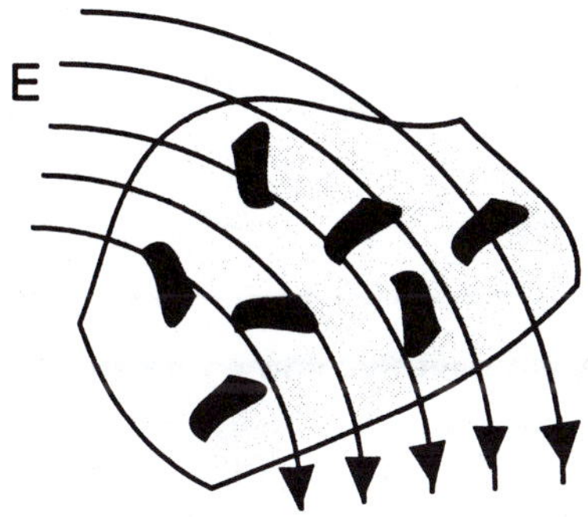

Fig. 2 A dielectric composite with multiple inclusions.

If now we assume that $\bar{\bar{\varepsilon}}_{eff}$ is the effective permittivity of the composite under consideration, then one can write the average displacement

$$D_{av} = \bar{\bar{\varepsilon}}_{eff} E_{av} \tag{7}$$

and

$$D_{av} = \varepsilon_{host} E_{av} + \langle \bar{P} \rangle \tag{8}$$

where the polarization due to each inclusion is

$$\bar{P} = \bar{\bar{\alpha}} \cdot E_{\ell} \tag{9}$$

$\bar{\bar{\alpha}}$ is the numerically found polarizability.

Combining the above equation one can write the following general expression for $\bar{\bar{\varepsilon}}_{eff}$,

$$\bar{\bar{\varepsilon}}_{eff} = \varepsilon_{host}\,\bar{\bar{I}} + \left(\bar{\bar{I}} - \frac{1}{\varepsilon_a}\int_{4\pi} d\Omega\,\omega_0(\Omega)\bar{\bar{\alpha}}\cdot\bar{\bar{L}}\right)^{-1}\cdot\int_{4\pi} d\Omega\,\omega_0(\Omega)\bar{\bar{\alpha}} \tag{10}$$

where $\bar{\bar{I}}$ is the unit dyadic.

In our calculation we have assumed that composite structure under consideration is excited with uniform electric field. Also, the randomly oriented inclusions are uniformly distributed throughout the composite structure. These assumptions simplify the above effective permittivity equation [10]. The apparent permittivity ε_a has been calculated using the following relation [9,10]

$$\varepsilon_a = \varepsilon_{host} + a\left(\frac{1}{3}\mathrm{tr}[\bar{\bar{\varepsilon}}_{eff}] - \varepsilon_{host}\right) \tag{11}$$

where $\mathrm{tr}[\bar{\bar{\varepsilon}}_{ff}] = \varepsilon_{xx}+\varepsilon_{yy}+\varepsilon_{zz}$. Parameter "a" can be selected to match the experimental results. The benefit of this approach is that by the proper choice of parameter "a", other phenomena such as effects of the screening charges at the grain boundaries can be included in the model.

The polarizability $\bar{\bar{\alpha}}$ can be calculated in closed form for certain shapes such as sphere or ellipsoid but in general it should be numerically calculated. We have numerically calculated $\bar{\bar{\alpha}}$ for spheroid, ellipsoid of revolution and cubic dielectric inclusions. We start with the system shown in fig. 3, where a dielectric body is immersed in a uniform electric field E_0. The normalized dipole moment per unit volume as a function of the inclusion's permittivity is shown in figs. 4 and 5. Our results were presented along with the values obtained using analytical methods. As shown in the figure, for spheroid and cubic dielectrics with $\theta=0$, the dipole moment is along Z-axis. For both cases, the polarizability is isotropic. In addition, we observe that cubic inclusion results in smaller dipole moments then spheroid inclusion of the same volume.

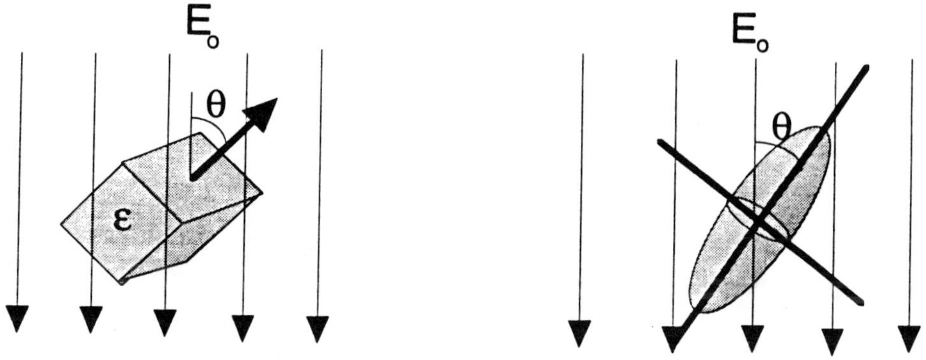

Fig. 3 Simulated arrangement for numerical calculation of the polarizability of a single inclusion.

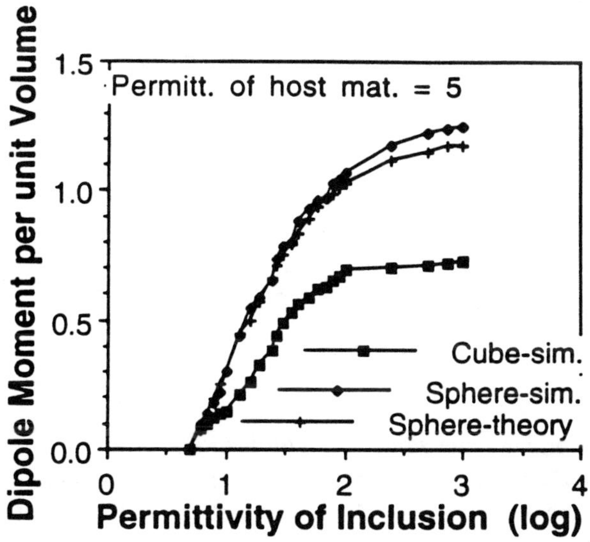

Fig. 4 Normalized dipole moment $\left[\frac{P_z}{4\pi\varepsilon_0 E_0}\right]$ per unit volume for cubic and spherical inclusions, as a function of the inclusion's permittivity.

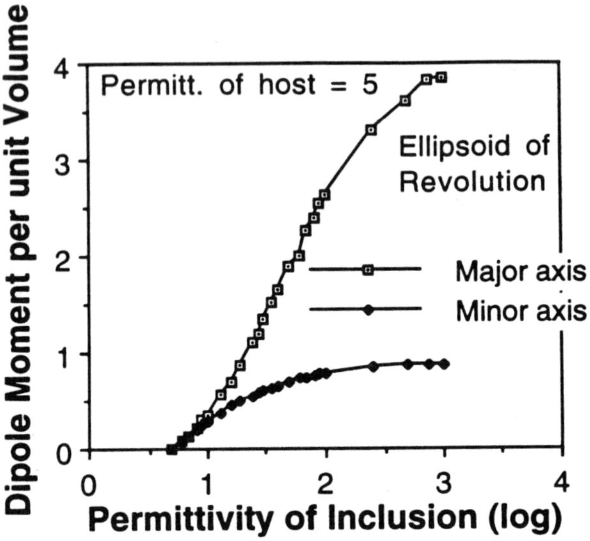

Fig. 5 Normalized dipole moment $\left[\frac{P_i}{4\pi\varepsilon_0 E_0}\right]$ per unit volume for an ellipsoid inclusion with 3/1 axial ratio.

For our calculation we have selected an ellipsoid of revolution with an axial ratio of 3. Results indicate that, as expected, the dipole moments are different along the major and the minor axes. Also, the polarizability tensor depends on the ellipsoid's axial ratio.

Fig. 6 shows the calculated effective permittivity of the composite structure with different shape inclusions. It is interesting to note that the shape of the inclusion significantly changes the permittivity value.

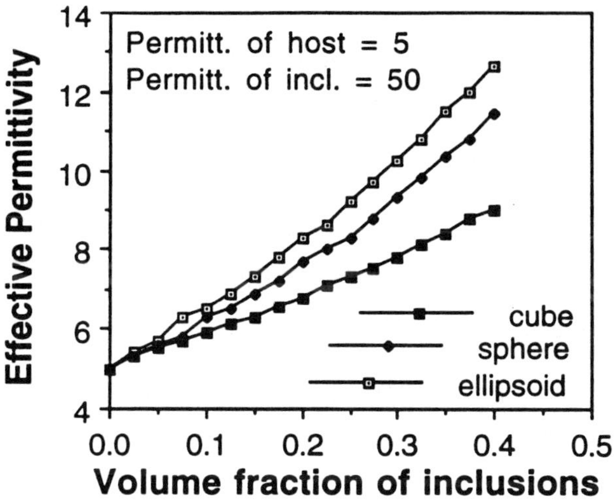

Fig. 6 Calculated effective permittivity of mixture as a function of volume fraction of the inclusions.

In conclusion, we have presented here a numerical solution using the boundary integral method. This approach provides an accurate and user friendly computer software which can be used for design and analysis of composite ceramics. In addition, this method can also be extended to predict the effects of various imperfections.

REFERENCES

[1] J.C. Maxwell-Garnett, "Colours in metal glasses and in metal films," Philos., R. Soc., London, Vol. 203, pp. 385-420, (1904).

[2] L.K.H. VanBeek, "Dielectric behavior of heterogeneous systems," Progress in Dielectrics, Lond., Vol. 7, pp. 76-77, 92-101, (1967).

[3] L.E. Nielson, "The permittivity of suspensions and two phase mixtures," J. Phys., D: Appl. Phys., Vol. 7, pp. 1549-1555, (1974).

[4] A.H. Sihvola, "Self consistency aspects of dielectric mixing theories," IEEE Trans. GeoSci. and Remote Sen., Vol. 27, No. 4, pp. 403-415, (1989).

[5] T.B. Lewis and L.E. Nielsen, "Dynamic mechanical properties of particulate filled composites," J. Appl. Polm., Sci., 14, pp. 1449-1471, (1970).

[6] P.K. Ghosh, M.E. Azimi, "Numerical calculation of effective permittivity of lossless dielectric mixture using boundary integral method," IEEE Trans. on Dielec. and Elect. Insu., (in press).

[7] R.F. Harrington, Field computation by moment methods," Krieger, Melbourne, Florida, (1982).

[8] A. Yaghjian, "Electric dyadic green's functions in the source region," Proc. IEEE, Vol. 68, No. 2, pp. 248-263, (1980).

[9] A.H. Sihvola and J.A. Kong, "Effective permittivity of dielectric mixtures," IEEE Trans. on GeoSci. and Remote Sens., Vol. 26, No. 4, pp. 420-429, (1988).

[10] M.E. Azimi, P.K. Ghosh, "A numerical approach to analyze the internal electrical response of dielectric ceramics with arbitrary shaped inclusion," ISAF '94 meeting, Penn State University, Univ. Park, Aug. 7-10, (1994).

Scientific, Regulatory and Socioeconomic Analysis of Inorganic Lead Compounds
Focus on Electroceramics and Glasses

Nathan Nair, Amar Bhalla and Rustum Roy
The Pennsylvania State University, University Park, PA 16802

ABSTRACT

Inorganic lead is not an essential element required for normal physiological activity. It is a cumulative toxin that can cause significant tissue damage depending on the blood and tissue concentrations. "High technology" products containing lead include electroceramics and glasses. "Low technology" products containing lead include paints, piping and solders. Although many electroceramics and glasses have greater lead by weight than some of the "low technology" materials, the "high technology" materials have less significant environmental effects. Until replacement materials are developed, these "high technology" products can not be regulated in the same manner as the "low technology" materials.

INTRODUCTION

Lead based compounds are numerous. The reigning public view of lead based compounds groups them all under the title "Lead" and does not draw distinctions between the separate compounds. However, the idea deteriorates under closer examination. When the toxicological effects of lead compounds are studied, a distinction between organic and inorganic lead compounds appears since each group behaves differently in biological systems[1]. Further examination of inorganic lead compounds exposes additional differences in the stability, modes of human exposure and socioeconomic importance of these compounds. Lead based electroceramics and lead glasses are two families of inorganic lead compounds that may be controlled in the future when restrictions on use of lead are enacted. These compounds are widely used in electronics, yet pose minimal health risks due to their stability. The socioeconomic impact of restricting these inorganic lead compounds could be significant because of their widespread use and the difficulty of locating replacement materials. This paper will therefore compare lead based electroceramics and lead glasses with other inorganic lead compounds to determine whether all of these should be restricted.

To the extent authorized under the laws of the United States of America, all copyright interests in this publication are the property of The American Ceramic Society. Any duplication, reproduction, or republication of this publication or any part thereof, without the express written consent of The American Ceramic Society or fee paid to the Copyright Clearance Center, is prohibited.

The focus of this paper will be a comparison between inorganic lead compounds. Thus organic lead compounds, such as tetramethyllead and tetraethyllead, will not be considered in this discussion. Information that characterizes the effects of lead on the human body defines the seriousness of the problem with all inorganic lead compounds. Descriptions of stability and modes of human exposure for different compounds demonstrate which compounds pose the most serious danger and should be regulated. The present regulatory status and future socioeconomic impacts of a ban are critical factors to be considered before adopting additional regulations. This report is formally divided into the following sections:

(1) Environmental and Health Issues
(2) Stability of the Compounds
(3) Modes of Human Exposure
(4) Federal Regulations
(5) Socioeconomic Effects

ENVIRONMENTAL AND HEALTH ISSUES

Environmental and human health effects are intricately linked. As inorganic lead compounds enter the environment, these substances will also enter the biological chain. Once within the biological chain, inorganic lead will eventually be absorbed by humans. Human exposure generates additional areas of concern, (1) where the inorganic lead is retained in the body and (2) what are the effects of lead on organ. These issues and the possible lead sources differ for different nations and geographic regions.

Environmental

Inorganic lead enters the environment from numerous sources. Natural sources of lead include volcanic debris and the erosion of certain soils and clays. Human sources of lead atmospheric pollution include industrial processes such as smelting and lead battery manufacturing. Cigarette smoke also contains lead compounds[1]. The atmospheric lead concentration adjacent to smelters may rise to as high as 93 ($\mu g/m^3$) [2]. These sources will significantly effect only the local atmosphere. Although not a consideration of this paper, the largest general source of lead in the atmosphere has been the combustion of organic lead containing gasoline. Before the leaded gasoline ban, American cities had an average level of 0.89 $\mu g/m^3$ and average levels in rural areas was 0.11 $\mu g/m^3$ in 1974[3]. Except for those individuals who live or work near sources, population exposure to atmospheric lead is minor compared to subsequent pathways. The major significance is those heavy metals, including lead, have a tendency to concentrate at the lower atmospheric levels and contaminate surrounding soil layers[4]. Soil lead contamination is in certain circumstances a direct effect of atmospheric pollution. Lead may be stabilized by the soil humus, or its organic component, and remain in the upper 2.5 centimeters of soil, unless it is disturbed. In these cases the lead may move deeper into the soil layers. Ground water supplies can be contaminated by municipal or hazardous dump sites[4].

Absorption

Inhalation and ingestion are the two routes for human exposure to lead. For adults, inhalation is the more efficient mode of absorption, with a 30-37% absorption rate into the blood stream through our respiratory tract. Ingestion, in adults, will lead only to 10% of lead entering the blood stream. However, for children ingestion may cause up to 50% absorption of lead into the blood stream[5].

Table 1. Environmental Lead Intake and Absorption[7]

Source of Lead Exposure	Percentage of Absorption (μg)
Food and Beverages 10-30 μg/liter	5-15
U.S. City Air 1-6 $\mu g/m^3$	25-50
Cigarette Smoke 1 μg/cigarette	10-15
Industrial air 200 $\mu g/m^3$	15

Lead in water is absorbed at a greater rate, 35-50% for adults and greater than 50% for children[6]. Although inhalation of lead has a greater absorption rate, ingestion is the more significant route of human exposure for those who are not exposed to direct sources of atmospheric lead pollution.

Inhalative absorption occurs in the alveoli, or microscopic air sacs, of our lungs where blood gases are exchanged with atmospheric gases. Analysis of the air lead versus blood lead levels is described as generally curvilinear. As air lead levels arithmetically increase, the blood level concentration grows at an increasing rate[8]. The derivative of this function for males is greater than for females. The derivative is also higher for children as compared to adults[9]. Adults, without prior lead exposure, will experience a 1 μg/dL increase in blood lead concentration for every 1 $\mu g/m^3$ increase in air lead concentration. For children, this simplified relationship may be a 2 μg/dL or greater increase in blood lead for a 1 $\mu g/m^3$ air lead increase[10].

The ingestive absorption of lead from food and beverages is variable. The lead concentration of foods and beverages may increase after handling, processing and storage so that it is significantly greater than the raw food lead concentration. Blood absorption of lead from food varies with consumer age, sex and eating habits. Children may be further exposed to lead while ingesting soil and dust. For children who live near smelters, this accounts for a major source of lead absorption. When soil lead concentrations reach 500 parts per million, a 2 μg/dL increase in blood lead levels of children occurs for every 100 ppm increase in soil concentration. However since 1980, the average lead intake of adolescents, toddlers, and infants has decreased as shown in Figure 1.

Health Effects

Certain metals are required for normal physiological activity. Calcium, potassium, sodium, magnesium, and iron range from 0.1-1.5% of body mass. Other essential metals include chromium, cobalt, copper, manganese, molybdenum, selenium, tin, vanadium, and zinc are required in trace amounts. Lead is not required for regular activity. It is a cumulative poison that is retained by tissues and can cause acute and chronic illnesses. Blood, soft tissues and skeletal tissues each have a different normal lead level and different retention rates. Once lead exposure reaches significant levels, dose-effect correlations are used to establish the human health effects. Individual dosages of lead are related to specific human effects. Blood lead levels will be the scale used to establish the dosage. Dose-effect studies are specified for three primary organs, (1) blood, specifically heme synthesis, (2) central nervous system and (3) the kidneys.

Table 2. Lead Content of Unexposed Adult Human Tissues[7]

Tissue	Lead Content
Blood	11-38 μg/dL
Soft Tissues	mg/100g Wet Tissue
Liver	0.40 - 0.28
Lung	8×10^{-5} - 0.03
Hilar lymph nodes	4×10^{-6} - 0.04
Kidney	0.05 - 0.16
Spleen	0.01 - 0.07
Brain	0.01 - 0.08
Skeletal Tissue	mg/100g Wet Tissue
Bone	0.67 - 3.59
Teeth	0.007 ± 1.170

Normal blood lead concentrations vary considerably among individuals. Geographic location, occupation and living habits may effect the blood lead concentration. A suburban, nonsmoker from Philadelphia may have a blood lead level of 11 μg/dL, while a garage mechanic from Cincinnati may have a blood lead level of 38 μg/dL[7]. As discussed earlier, lead enters blood from inhalation, ingestion, or diffusion from lead stored in bones or soft organs. During the early stages of lead exposure, blood lead levels will correspond to the concentrations of recently inhaled or ingested lead[11]. For individuals who have had long term lead exposure, blood lead levels will correspond to the amount of lead stored in bones or soft organs[12]. Ninety percent of blood lead is bound to erythrocytes, or red blood cells, where it has a half-life of 36±5 days[13]. The remaining 10% of blood lead is in plasma, or the liquid portion of blood, where the lead has a half-life of minutes or hours[14]. Lead will leave blood mainly from the plasma portion and is excreted in urine. The average lead levels of some soft organ tissues are located in Table 2. The half-life of lead in soft tissue is 40 days. A percentage of this lead is rapidly exchanged with blood plasma. Lead is excreted from soft tissues into gastrointestinal secretions, sweat, hair and nails[14]. The normal lead levels of skeletal tissues, including bone and teeth, are located in Table 2. 99% of lead absorbed from the environment will eventually deposit in skeletal tissues[15]. The half-life of skeletal lead is 10000 days and can be considered inactive once deposited.

Anemia, or the lack of hemoglobin, is a major symptom of lead exposure. Hemoglobin is an iron based bio-molecule that acts as the primary mode of oxygen transport from our lungs to our tissues. Multiple studies have shown that increased lead concentrations in the blood stream correspond to decreased hemoglobin levels[16,17]. Blood lead levels as low as 5 μg/dL have been shown to cause subtle decreased hemoglobin levels[18]. Frank anemia in adults has been observed at lead levels as low as 50 μg/dL. Children appear to be the most highly effected age group, displaying less tolerance for blood lead[19]. Lead inhibits the enzymatic or catalytic activity of two molecules that are necessary for hemoglobin production. The primary enzyme is δ-aminolevulinic acid dehydratase (ALAD) of the red blood cells[20]. In addition to preventing ALAD activity in the red blood cells, in children ALAD of the brain is also inhibited by lead. The second enzyme effected is ferrochelatase. Ferrochelatase catalyzes the reaction between ferritin and protoporphyrin to form heme, a component in hemoglobin. Ferrochelatase is inhibited by lead, causing an excess of another bio-molecule, protoporphyrin, in red blood cells. At blood lead levels of 15 μg/dL, increased levels of protoporphyrin may be noted.

Acute lead encephalopathy, or alteration of the brain structure, has been reported at lead levels above 120 μg/dL in adults[21,22]. Children again appear to be at greater danger, with levels as low as 80 μg/dL causing acute encephalopathy[10]. Lower lead levels may also cause neurological problems. Children who do not exhibit symptoms of acute encephalopathy but have elevated dentine lead levels, have on average 4.5 points lower IQ verbal examination scores than other students with lower dentine lead levels[23]. Asymptomatic adults with elevated blood lead levels may exhibit memory, attention or concentration difficulties[14]. The hypothesized mechanism for lead neuropathy, or subtle brain alteration, is attributed to damaged endothelial cells of blood capillaries in the brain. Lead causes these cells to expand, which increases the permeability of the blood-brain barrier to other substances that can cause acute encephalopathy

vessels[24]. This may explain why suckling infants are at greater risk than the mother who is directly intaking lead. Figure 2 demonstrates that children with high lead levels have lower verbal intelligence quotients than their counterparts with lower lead levels. From the figure, 50% of children with low lead levels have verbal IQ scores lower than approximately 100 points. However, 50% of children with high lead levels have verbal IQ scores lower than approximately 90 points.

One of the major functions of the kidneys is nitrogenous waste removal. Nitrogenous wastes are formed during the metabolism of proteins. Lead has been shown to effect the proximal convoluted tubules of kidneys[15]. The primary function of these proximal tubules is to reabsorb water, sodium ions and amino acids back into the bloodstream so that these metabolically important chemicals are not excreted in urine. Lead inhibits the enzyme 1-a-25-hydroxyvitamin D hydroxylase of the cells lining the inner wall of the proximal tubules. This reduces the ability of the tubules to reabsorption amino acids, leading to the aminoaciduria or acidic urine[14]. Long-term, high dose exposure to lead will eventually cause partial renal fibrosis, which will result in irreversible reduction of kidney function and finally chronic renal failure[25].

STABILITY

Table 4 includes data concerning the solubility of several inorganic lead compounds. Solubility data can be used to determine whether a compound posed the risk of causing environmental lead contamination. This data can be compared with lead based electroceramics and lead based glasses.

Two significant lead based pigments used in paints are lead carbonate (basic white lead) and lead oxide (red lead). Both compounds are soluble under acidic conditions as listed in Table 4. The solubility of lead solders is similar to that of metallic lead, which will dissolve in slightly acidic pH. Lead sulfate ($PbSO_4$), used in storage batteries, is soluble in water in either neutral or acidic pH[26]. Since the pH of rain and ground water is slightly acidic, they will cause the above compounds to dissolve and enter the environment.

Lead based electroceramics, including lead titanate (PT), lead zirconium titanate (PZT), lead lanthinum zirconium titanate (PLZT) and lead magnesium niobiate (PMN), are water insoluble. The leachablity of lead ions from these compounds is also minimal because of their complex crystal structure. Lead based glasses, including lead aluminum silicates, lead aluminum boro-silicates, and lead boro-composites, are also water insoluble and leachability is minimal due to their complex glass structure.

Table 4. Solubility (100 ml) of Inorganic Lead Compounds[26]

Material	Cold water	Hot water	Acid	Base
Metallic Lead	Insoluble	Insoluble	Nitric and Acetic Acid	Insoluble
Pb-Sn Soft Solder	Insoluble	Insoluble	Nitric and Acetic Acid	Insoluble
Lead arsenate	Insoluble	Slightly Soluble	Soluble	Soluble
Lead carbonate	Insoluble	Insoluble	Soluble	Soluble
Lead chromate	Slightly Soluble	Slightly Soluble	Soluble	Soluble
Lead oxide	Insoluble	Insoluble	Soluble	Insoluble
Lead silicate	Insoluble	Insoluble	Decompose	Soluble

MODES OF HUMAN EXPOSURE TO LEAD

Presently, the two major sources of human exposure to inorganic lead are from (1) lead based paints and (2) lead piping used in many public water distribution systems. These sources of exposure should then be compared to (3) lead based electroceramics and lead glasses.

Lead Based Paints

Children are at the most significant risk for lead exposure from lead based paints. Fourteen million housing units in the United States contain lead paint that is deteriorating and of these 3.8 million are the homes of young children[4]. Children in these conditions are susceptible to lead exposure through ingestion of lead contaminated dust particles. Children prone to pica, or the regular ingestion of non-food materials, are also at risk if paint chips are consumed. Home renovation of houses using lead paint, which includes most homes constructed before 1980, is another possible point of exposure for children[4]. In addition, children are at further risk because of their lower tolerance for lead.

Lead Piping

Drinking water is another possible source of human exposure. As stated earlier, lead levels in ground water are typically low unless it is near municipal dump sites, or industrial producers of atmospheric lead. However, public water distribution systems may contain lead connectors, lead piping, and lead soldered joints for copper piping (lead-tin alloy). Housing units built before 1920 contain mostly lead piping. After 1950, lead piping was generally replaced by copper piping that used lead solder. Cisterns for water storage or collection are another possible source of lead when lead solder is used in their construction or for repairs[4]. The probability of significant exposure is greater if the plumbing is only used intermittently and the system is not flushed prior to use.

Lead Based Electroceramics and Lead Based Glasses

Lead based electroceramics have multiple applications in electronics. These uses include capacitors, transducers, sensors, phase shifters, pyro-detectors, thermistors, memory devices, and medical diagnostic applications. Lead based glasses are major ingredients in display sealing, integrated circuit packages and various coatings. These electronics components are incorporated into cardiac pacemakers, analog devices, and telecommunication systems and various consumer electronic systems[27,28]. Worker exposure during production is minimized through proper handling, protection and efficient production methods. During product use, humans are exposed to minimal amounts of lead since the actual lead components are inside the consumer product. Certain medical products, such as pacemakers or transducers, will significantly increase the risk of human exposure. However, in these situations the product use is necessary and the risk of human exposure is warranted.

Production and Disposal

Industrial disposal of waste materials from production of the lead containing compounds is another source of human exposure. The production of lead based compounds will produce waste materials that require disposal. Processing steps, including melting, preparing "slurries," and mixing powders, can cause additional lead contamination of the environment. Finishing steps, including cutting, polishing, and molding, will create waste materials that are not used in the final products. Lastly, disposal of lead containing components after their life span will further contaminate the environment.

FEDERAL REGULATIONS

Table 5. Standards and Regulations for Lead[29]

Agency	Focus	Level of Lead
CDC	Blood	25μg/dL-level of concern
OSHA	Air	50μg/m^3
OSHA	Blood	60μg/dL-removal from exposure
ACGIH	Air	150μg/m^3
EPA	Air	1.5μg/m^3-3 month average
EPA	Water	50μg/L
FDA	Food	100μg/day
CPSC	Paint	0.06% by weight

Lead based paints containing greater than 0.06% of lead by weight were banned by the Consumer Product Safety Commission in 1978 for consumer products or for residential purposes under Federal Hazardous Substances Act. Before 1950, paints may have contained nearly 50% of lead by weight[30]. Exceptions were given to paints used for industrial, military or marine uses[4]. Although the Act prohibits future use of lead based paints, 74% of existing housing units currently contain approximately 3 million tons of lead in the paint used. Figure 3 demonstrates how acceptable lead levels set by federal agencies has decreased since 1975. Presently, there is no standard for safe lead levels for existing painted surfaces such as housing units. Most existing standards are based primarily on practical issues, instead of health concerns[4]. In 1992, the Lead-based Paint Exposure Reduction amendment to the Toxic Substances Control Act was passed through Congress[31]. The amendment required the Department of Health and Human Services and Department of Housing and Urban Development to develop regulations concerning lead based paint activities, including training programs and home renovation[31]. Drinking water safety is another issue that the federal government has regulated. The 1986 Safe Drinking Water Act banned lead in water distribution systems, and it limited the percentage of lead in brass plumbing to 8%[4].

In 1993, a bill was proposed in the United States Senate to create comprehensive regulations concerning a wide range of lead containing compounds. The Lead Reduction Exposure bill was passed by the Senate, but the House of Representatives did not pass this amendment to the Toxic Substance Control Act before the session ended. If passed in a later session of Congress, it would place restrictions on lead solders, lead paints, lead piping. It may also restrict the manufacture of lead based electroceramics and lead based glasses because of their relatively high lead content by weight[32].

SOCIOECONOMIC EFFECTS OF LEAD REPLACEMENT

The following criteria will be considered in this section: (1) The range of uses and (2) the practicality of replacement. These criteria will be applied to a range of inorganic lead compounds. The importance of these concerns of replacement varies for industrialized and developing nations and this is another consideration.

Range of Uses

Lead oxide red and lead carbonate are pigments used in lead-based paints. Until 1950, lead based paints contained 50% lead by weight. Use of these as interior paints declined until 1970, when Consumer Product Safety Commission banned the use of lead based paints containing more than 0.06% lead by weight in 1978 for most uses. Presently, lead based paints can be used for industrial, military, and marine uses[4]. The production of lead sulfate for storage batteries accounts for the largest single use of lead[7]. Although lead storage batteries are widely used because of their convenience, the cost of these sources of energy is approximately $80 per kilowatt-hour. However, power plants only charge approximately $0.10 per kilowatt hour[33].

Batteries use lead sulfate and hydrosulfuric acid because the batteries can be recharged after the reduction-oxidation reaction between lead (Pb) and lead oxide (PbO_2) has converted to lead sulfate[33]. Metallic lead and lead alloys are used in lead piping and solder. Both are still widely used in older buildings and older public water distribution systems. A major similarity between all of these compounds is that they each have very specific, usually "low technology," uses.

Lead based electroceramics and lead based glasses are multi-use compounds. As stated earlier, these compounds are used in electronics including capacitors, transducers, sensors, phase shifters, pyro-detectors, thermistors, memory devices, medical diagnostic applications. Lead based glasses are a major ingredient in display sealing, integrated circuit packages and various coatings. These electronic components are incorporated into cardiac pacemakers, analog devices, and telecommunication systems and various consumers electronic systems. In general, these compounds are widely used, in "high-technology" applications.

Replacement

Lead-based paints have already been limited by the United States federal government. Non-lead containing pigments can be used to effectively replace lead based pigments without excessive expense. Storage batteries are more difficult to replace because an oxidation-reduction reaction that produces one compound is necessary for rechargeable batteries. Because of the expense and inefficiency of lead storage batteries, research into other storage batteries has continued, especially in connection with photo-voltaic cells. Nickel-iron batteries are lighter than lead storage batteries and are presently used in telephone exchanges and in other heavy duty situations. Nickel-cadmium batteries are used in spacecraft and produce 1.3 volts as compared to a lead storage battery that can produce 2 volts. The Nickel-cadmium batteries may also last up to twenty-five years. However, this may not be an alternative due to the extreme toxicity of cadmium. Lead solder and lead piping, used in water distribution systems, are being replaced by copper, brass or poly (vinyl chloride) piping. Lead solders used in food containers are also being replaced as demonstrated by Figure 4.

Replacement materials for lead based electroceramics and lead glasses that have equivalent efficiency are not presently being used or phased into use. Research is being conducted to produce suitable replacement materials. However, unlike the other described compounds, replacements for electroceramics and glasses must meet a higher or equivalent standard of efficiency and should be inexpensive. Lower quality or more expensive compounds could not effectively be phased into electronic devices that presently use electroceramics and glasses without creating economic problems.

Changes in production and disposal methods may be an alternative to replacement. Low temperature processing of lead based compounds will reduce the vapor pressure and the amount of lead lost from vaporization. Clean cutting methods with lasers will reduce the amount of material wasted. If possible, avoiding cutting and polishing can significantly reduce the amount of waste material containing lead. Proper packaging of final components may reduce environmental contamination at the end of its life span.

The Concerns of Industrial versus Developing Nations

To defend their technological competitiveness, industrialized nations will continue to use and develop electroceramics and glasses that contain lead compounds. However, these nations are also more able to replace "low technology" products containing lead, including lead paints, lead piping, and leaded gasoline. Thus production and disposal of high technology products containing lead will remain the major source of lead pollution. Developing nations will be concerned with the disposal of these high technology products, rather than production. The major concern of these nations will be the "low technology" products containing lead, including lead paints, lead piping and leaded gasoline.

CONCLUSION

Three basic criteria should be used to determine whether a compound is sufficiently hazardous that it should be restricted. The primary criteria are the human health effects of a compound. Certain compounds will inflict serious health problems after minimal, short term contact. Inorganic lead is a cumulative poison that will cause widespread health effects after long term exposure to significant doses. Once in the human body, all inorganic lead compounds act similarly. However, proper lead concentration monitoring and medical treatment will avert serious effects in almost all situations.

The second criterion is the stability of the compound and at what concentrations humans will be exposed. Unstable compounds that will readily enter the environment pose a more serious threat to the community than compounds that remain inert. An additional aspect of this criterion is the relative human exposure to these substances. Compounds, such as lead oxide (red lead), lead carbonate, metallic lead and lead alloys, will readily enter the environment and humans are widely exposed to consumer products that contain these inorganic compounds. However, lead based electroceramics and lead glasses will not as readily enter the environment and human exposure is variable. When used in electronic components, human exposure is minimal. When used in medical applications, exposure rises but is offset by the necessity of the product and stability of the compounds.

The third criterion is a cost of replacement versus benefits analysis. Inorganic lead compounds used in lead paints, batteries or lead piping already have available replacement materials in use. In addition, these compounds perform very narrow functions. The cost of new product development and implementation will have a lesser effect on society and the economy. However, lead based electroceramics and lead glasses are used widely as the basic components of many electronic devices. These compounds are not as readily replaced by similar quality materials. If effective replacement materials are not found, phasing out these compounds in the United States could have serious socioeconomic effects. Other nations that have not restricted the use of lead based ceramics and glasses will produce either higher quality or less expensive electronics. Even if society is willing to accept these consequences, a separate socioeconomic problem is created with medical applications. A more costly or less effective pacemaker or transducer will have a serious effect on the cost and quality of health care in the United States, In the present climate, neither of these is a viable option that American society will find acceptable.

These three criteria support the idea that certain inorganic lead compounds are extremely hazardous and warrant regulation. Most of these compounds have already been restricted or banned by the United States federal government after individual consideration. However, multilateral regulation of inorganic lead compounds cannot be supported by the presented data. The enormous differences in chemistry, uses and importance of inorganic lead compounds should be considered before the federal government acts. The federal government does have a responsibility to act on the prevailing public opinion if society is completely aware of the future consequences that may be involved. However, society has not been exposed to accurate information concerning the dangers of inorganic lead compounds and in this situation the federal government has two other responsibilities. The first is to correctly inform the public of the different types of lead compounds and their uses. The second responsibility is to act carefully and cautiously, rather than submit to a public opinion that is based on incomplete data.

Lead based electroceramics and lead glasses should not be placed in the hazardous materials category with the other described inorganic lead compounds, but the regulation of the raw material supply, production and disposal is desirable. Lead based electroceramics and glasses have too many important uses to be banned in an attempt to completely eliminate lead contamination and human exposure. This type of regulation would be counterproductive for our society and economy and may practically cause far more problems than it could solve. Careful, specific action against harmful compounds is a more reasonable solution than legislating entire element lead and all the compounds produced with it.

REFERENCES

[1]Agency for Toxic Substances and Disease Registry, Toxicological Profile for Lead. Agency for Toxic Substances and Disease Registry. Syracuse research Corporation (Public No. ATSDR/TP-88/17). Atlanta (1990).

[2]Landrigan, P.J., Gehibach, S.H., Rosenblum, B.F., Shoults, J.M., Candelaria, R.M., Barthel, W.F., Liddle, J.A., Smrek, A.L., Staehling, N.W. and Sanders, J.F., N.Engl. J. Med. 292. 123-129 (1975).

[3]Akland, G.B., Air Quality Data for Metals 1970 through 1974 from the National Air Surveillance Network. U.S. Environmental Protection Agency. EPA-600/4-76-041. Washington D.C (1976).

[4]Centers for Disease Control, Preventing Lead Poisoning in Young Children - A Statement by the Centers for Disease Control. Centers for Disease Control. Atlanta (1991).

[5]Alexander, F.W., Clayton, B.F. and Delves, H.T., In: Environmental health aspects of lead. Commission of European Communities. Luxembourg. pp. 319-331 (1972).

Table 3. Summary of lowest lead-induced health effects in adults[1]

Lowest Observed Lead Level (μg/dL)	Heme Synthesis and hematological effects	Neurological Effects	Cardiovascular Effects	Kidney Effects	Reproductive Effects
100-120		Symptoms of Encephalopathy		Chronic Nephropathy	
80	Frank Anemia			↓	
60		↑ ↑ ↑		↓ ↓ ↓	Female Reproductive Effects
50	Reduced Hemoglobin production	Overt subencephalopathic symptoms		↓ ↓ ↓	Alter Testicular function
	increased urinary ALAD and coproporphyrins	↓ Slowed nerve conduction ↓		↓ ↓	↓ ↓ ↓
30		↓ ↓	Elevated Blood Pressure (White Males, aged 40-59)		
25-30	Increased Erythrocyte Protoporphyrin (EP) in males		↓ ↓ ↓ ↓		
15-20	Increased EP in females		↓ ↓		
<10	ALAD inhibition		?		

or more subtle alterations. Newly developing capillaries appear more vulnerable to the effects of lead than already formed

REFERENCES

[1]Agency for Toxic Substances and Disease Registry, Toxicological Profile for Lead. Agency for Toxic Substances and Disease Registry. Syracuse research Corporation (Public No. ATSDR/TP-88/17). Atlanta (1990).

[2]Landrigan, P.J., Gehibach, S.H., Rosenblum, B.F., Shoults, J.M., Candelaria, R.M., Barthel, W.F., Liddle, J.A., Smrek, A.L., Staehling, N.W. and Sanders, J.F., N.Engl. J. Med. 292. 123-129 (1975). "Epidemic Lead Absorption near ore smeltor and the rate of particulate lead"

[3]Akland, G.B., Air Quality Data for Metals 1970 through 1974 from the National Air Surveillance Network. U.S. Environmental Protection Agency. EPA-600/4-76-041. Washington D.C (1976).

[4]Centers for Disease Control, Preventing Lead Poisoning in Young Children - A Statement by the Centers for Disease Control. Centers for Disease Control. Atlanta (1991).

[5]Alexander, F.W., Clayton, B.F. and Delves, H.T., In: Environmental health aspects of lead. Commission of European Communities. Luxembourg. pp. 319-331 (1972).

[6]Agency for Toxic Substances and Disease Registry, The nature and extent of lead poisoning in children in the United States- A report to Congress. Agency for Toxic Substances and Disease Registry. Atlanta (1988).

[7]Clayton, G.D. and Clayton, F.E., Patty's Industrial Hygeine and Toxiciology, Vol 2A. John Waley and Sons Inc. New York. pp. 1687-1727 (1981).

[8]Chamberlain, A.C., Effect of Airborne Lead on Blood Lead. Atmos. Environ. 17. 677-692 (1983).

[9]Charney, E., Sayre, J. and Coulter, M., Pediatrics 65. 226-231 (1980). "Increased lead absorption in inner city children and where does the lead come from?"

[10]United States Environmental Protection Agency, Air Quality Criteria for Lead. U.S. Environmental Protection Agency. Office of Research and Development (Public No. EPA-600/8-77-017). Washington D.C (1977).

[11]Landrigan, P.J., Baker, E.L. Jr., Feldman, R.G., Cox, D.H., Eden, K.V., Oenstein, W.A., Mather, J.A., Yankel, A.J., and VonLindern, I.H., J. Pediatr. 89. 904-910 (1976). "Increaded lead absorption with anemia and slowed nerve conduction in children near a smeltor"

[12]Baker, E.L., Landrigan, P.J., Barbour, A.G., Cox, D.H., Folland, D.S., Ligo, R.N. and Throckmorton, J., Br. J. Ind. Med. 36. 314-322 (1979). "Occupational lead poisoning in the US: clinical and biochemical findingd related to blood lead levels"

[13]Rabinowitz, M.B., Wetherill, G.W. and Kopple, J.D., J. Clin. Invest. 58. 260-270 (1976). "Kinetic analysis of lead metabolism in healthy humans"

[14]Mahaffey, K.R., In: Dietary and environmental lead: human health effects. National Institute for Occupational Safety and Health. Cincinnati. pp. 421-445 (1985).

[15]Goyer, R.A. and Rhyne, B., Pathological Effects of Lead. Int. Rev. Exp. Pathol. 12. 1-121 (1973).

[16]Roels, H.A., Lauwerys, R., Buchet, J.P. and Urelust, M.T., Int. Arch. Arbeitsmed. 34. 97-108 (1975). "Response of free erythrocyte porphyrin and urinary delta-ALAD in men and women moderately exposed to lead"

[17]Grandjean, P., Br. J. Ind. Med. 36. 52-58 (1979). "Occupational lead exposure in Denmark:screening with the hematoflourometer"

[18]Roels, H.A., Buchet, J.P., Lauwerys. R., Hbermont, G., Bruaux, P., Claeys-Thoreau, F., Lafontaine, A. and Van Querscheide, J., Impact of Air Pollution by Lead on the Heme Biosynthetic Pathway in Scool-Age Children. Arch. Environ. Health 19. 310-316 (1976).

[19]Adebenojo, F.O., Clin. Pediatr. 13. 874-888 (1974). "Hemstologic status of urban black children in Philadelphia"

[20]Hernberg S., In: Singal, R.L. and Thomas, J.A. (Eds). Lead Toxicity. Urban and Schwarzenberg. Baltimore. pp. 367-399 (1980).

[21]Aub, J.C., Minot, A., Fairhall, L., Lawrence, T. and Resnikoff, P., Lead Poisoning. Williams and Wilkins Co. Baltimore (1925).

[22]Chisolm, J.J. Jr. and Harrison, H.E., The Exposure of Children to Lead. Pediatrics 18. 943-957 (1956).

[23]Needleman, H.L., Leviton, A. and Bellinger, D., New Engl. J. Med. 306. 367 (1982). "Lead-asociated Intellectual deficiency"

[24]Bradbury, Michael, In: The Concept of a Blood-Brain Barrier. John Wiley and Sons. Chicester (1994).

[25]Wedeen, R.P., Maesaka, J.K., Weiner, B., Lipat, G.A., Lyons, M.M., Vitale, L.F. and Joselow, M.M., Am. J. Med. 59. 630-641 (1975). "Occupational lead nephropathy"

[26]Patty, Frank A., Industrial Hygiene and Toxicology, Vol. 2. John Wiley and Sons Inc. New York. pp. 941-985 (1958).

[27]Hirano, S., Yogo, T., Kikuta, K., Kato, K., Sakamoto, W. and Ogasahara, S., Sol-Gel Processing and Characterization of Ferroelectric Films. Ceramics Transactions. 25. 19-32 (1992).

[28]McCandlish, C., Military and Medical Applications. Electronic Materials Handbook Volume 1(Packaging) pg 386 ASM International, Materials Park, Ohio (1989).

[29]United States Department of Health and Human Services, Case Studies in Environmental Medicine-Lead Toxicity. (1990).

[30]United States Department of Health and Human Services, Sixth Annual Report on Carcinogens. 240-243 (1991).

[31]United States Congress, The Lead-Based Paint Exposure Reduction Act. Public Law 102-550 (1992).

[32]United States Congress, Lead Reduction Exposure Act. Senate Bill 729. Sections 103-108 (1993).

[33]Brown, T.L., Le May, Jr, H.E. and Bursten, B.E., In: Chemistry-The Central Science. 5. Prentice Halls. Englewood Cliffs. pp. 721-726 (1991).

Figure 1-Average Daily Lead Intake

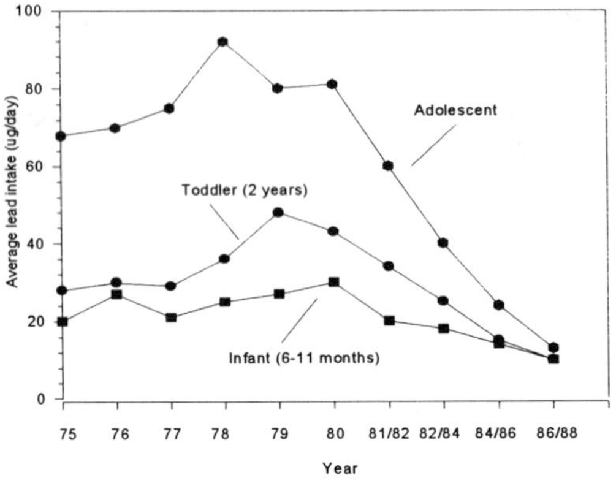

Source: U.S. Department of Health and Human Services, Public Health Service, Food and Drug Administration, Center for Food Safety and Applied Nutrition, FDA Total Diet Study (Washington D.C.:1989)

Figure 2-Comparison of Verbal IQ scores of children with High and Low Lead levels

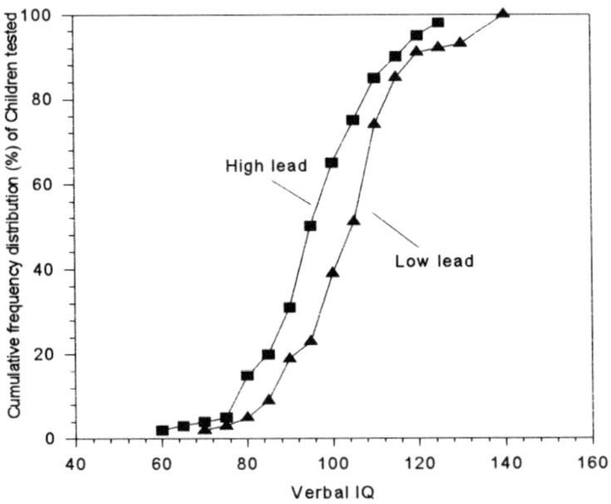

Source: Needleman, H.L, C. Gunnoe, A. Leviton, et al., "Deficits in Psychologic and Classroom Performance of Children With Elevated Dentine Lead Levels," New England Journal of Medicine 300. 689-695 (1979).

Figure 3-Changes in Blood lead Level Guidelines

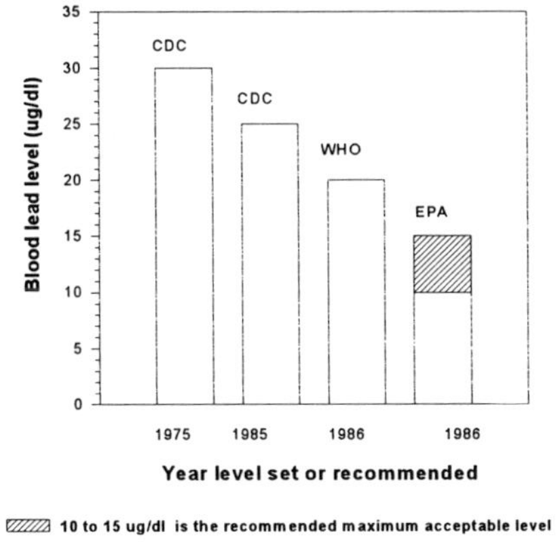

Source: U.S. Congress, Office of Technology Assessment, Neurotoxicity: Identifying and Controlling Poisons of the Nervous System, OTA-BA-436, 271 (1990).

Figure 4-Total Number of Cans Produced versus Lead Soldered Cans

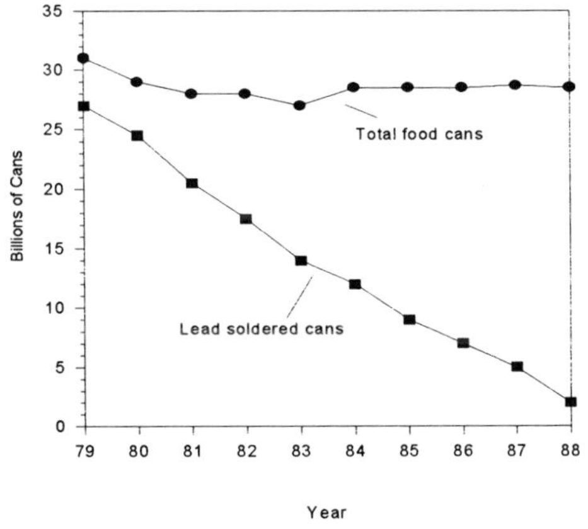

Source: U.S. Congress, Office of Technology Assessment, Neurotoxicity: Identifying and Controlling Poisons of the Nervous System, OTA-BA-436, 271 (1990).

KEYWORD AND AUTHOR INDEX